First published 1981 by CRC Press
Taylor & Francis Group
6000 Broken Sound Parkway NW, Suite 300
Boca Raton, FL 33487-2742

Reissued 2018 by CRC Press

Library of Congress Cataloging in Publication Data
Main entry under title:

east cell envelopes: biochemistry, biophysics,
and ultrastructure.

Jniscience series)
cludes bibliographies and indexes.
Yeast fungi. 2. Fungi—Cytology. I. Arnold,
ed Niels, 1936-. II. Title: Cell en-
es. [DNLM: 1. Yeasts—Cytology. 2. Cell
rane. QK 617.5 Y39]
.5.Y4 589.2'33 81-415
8493-5965-1 (v. 1) AACRl
8493-5966-X (v. 2)

of Congress record exists under LC control number: 81000415

Note
er has gone to great lengths to ensure the quality of this reprint but points out that some imperfections in the original copies
rent.

has made every effort to trace copyright holders and welcomes correspondence from those they have been unable to

315-89868-1 (hbk)
351-07778-1 (ebk)

Francis Web site at http://www.taylorandfrancis.com and the
te at http://www.crcpress.com

Yeast Cell Envelopes: Biochemistry, Biophysics, and Ultrastructure

Volume II

Editor

Wilfred Niels Arnold, Ph.D.
Professor
Department of Biochemistry
University of Kansas Medical Center
Kansas City, Kansas

CRC Press
Taylor & Francis Group
Boca Raton London New York

CRC Press is an imprint of the
Taylor & Francis Group, an informa bu

PREFACE

Yeasts hold important associations with man in at least three arenas. As industrial agents their positions in the fermentative and baking industries are household knowledge. A few yeast or yeast-like species are pathogenic. And, as experimental eukaryotic cells, yeasts continue their long history as useful subjects for studies on metabolism, genetics, and molecular biology.

A comprehensive review of the yeast cell envelope has not appeared previously and we trust that this attempt will be timely. The title of this volume was chosen to reflect the three major areas of contribution to our current understanding of the cell envelope, but we have not attempted to group chapters into subdivisions. That would be somewhat arbitrary at best. In fact, the contributing authors were recruited for their interdisciplinary work as well as their special expertise.

The approach is to describe phenomena, to review the literature, and to illuminate outstanding problems. We have also attempted to generate working hypotheses which may stimulate further studies. That some of these ideas be of germinal value is of more concern to us than that all of the hypotheses should stand the test of further experimentation.

Brenda Johnson has given special assistance in the assemblage of this volume. I also wish to acknowledge my former teachers and mentors Drs. G. Langdon, J. Middleton, J. Bald, and J. Thompson, as well as two colleagues, E. Juni and I. Goldstein, for encouragement during my formative years in biological chemistry.

W. N. Arnold
Westwood Hills, Kansas
July 1980

THE EDITOR

Wilfred N. Arnold, Ph.D., is Professor of Biochemistry at the University of Kansas Medical Center, Kansas City, Kansas. Dr. Arnold received a B.S. from the University of Queensland, Australia in 1956, an M.A. from the University of California, Los Angeles in 1958, and a Ph.D. from Cornell University, Ithaca, N.Y. in 1962. He has held postdoctoral fellowships at the Waite Institute, University of Adelaide, Australia and at the University of Wisconsin Medical School, Madison. Before coming to Kansas in 1971, he held research positions at the Commonwealth Scientific and Industrial Research Organization (CSIRO) in Australia and the University of California at Riverside; and an Assistant Professorship at Wayne State University Medical School, Detroit, Mich.

Dr. Arnold is a member of the American Society of Biological Chemists, American Chemical Society, American Association for the Advancement of Science, American Society for Microbiology, Society for Complex Carbohydrates, and the honorary society, Sigma Xi.

Dr. Arnold has authored 40 publications in biochemistry. His current research interests are centered on the yeast cell envelope with particular emphasis on questions at the interface of biochemistry and ultrastructure of pathogenic and nonpathogenic species of yeast. This research has been sponsored by N.I.H., Research Corporation, and Sigma Xi. Dr. Arnold has taught biochemistry to undergraduate, graduate, and medical students and has authored and directed three educational films.

CONTRIBUTORS

John S. D. Bacon, Sc.D., F.R.S.E.
Honorary Research Associate
University of Aberdeen and
Head
Carbohydrate Biochemistry
Department
Rowett Research Institute
Bucksburn, Aberdeen
Scotland, United Kingdom

Robert G. Garrison, Ph.D.
Research Microbiologist
Veterans Administration Medical
Center
Kansas City, Missouri and
Associate Professor
Department of Microbiology
University of Kansas Medical Center
Kansas City, Kansas

Jaroslav Horák, Ph.D.
Senior Worker
Department of Cell Physiology
Institute of Microbiology
Czechoslovak Academy of Sciences
Vídeňská, Prague
Czechoslovakia

Arnošt Kotyk, D.Sc.
Head
Department of Cell Physiology
Institute of Microbiology
Czechoslovak Academy of Sciences
and
Professor
Charles University
Vídeňská, Prague
Czechoslovakia

Oldřich Nečas, M.D., D.Sc.
Professor and Head
Department of Biology
Medical Faculty
J. E. Purkyně University
Brno, Czechoslovakia

Armando J. Parodi, Ph.D.
Visiting Scientist
Department of Microbiology
The Wellcome Research Laboratories
Research Triangle Park, North
Carolina

Martin L. Slater, Ph.D.
Executive Secretary
Microbial Physiology Study Section
Division of Research Grants
National Institutes of Health
Bethesda, Maryland

Eva Streiblová, Ph.D.
Senior Scientific Worker
Department of Cell Physiology
Institute of Microbiology
Czechoslovak Academy of Sciences
Vídeňská, Prague
Czechoslovakia

Shigeo Suzuki, Ph.D.
Professor
The Second Department of Hygienic
Chemistry
Tohoku College of Pharmacy
Komatsushima, Sendai
Japan

Augustin Svoboda, M.D., Ph.D.
Assistant Professor
Department of Biology
Medical Faculty
J. E. Purkyně University
Brno, Czechoslovakia

TABLE OF CONTENTS

Volume I

TABLE OF CONTENTS

Volume II

Chapter 1

ENZYMES

W. N. Arnold

TABLE OF CONTENTS

I. BACKGROUND

The yeast cell envelope contains an important complement of enzymes. Self-evident as that statement is now, it was rather the mechanical role of the cell wall that dominated earlier biological considerations. Consequently the envelope was looked upon as a relatively inert complex of integuments rather than an "outer metabolic region of the cell."[1]

Studies on the enzymology of the yeast cell envelope have their beginning in turn of the century work on isolation and partial purification of β-fructofuranosidase. By 1921 Willstätter's laboratory[2] had incidentally documented the peripheral location of β-fructofuranosidase, and other enzymes were subsequently associated with the cell envelope. These enzymes fall into two classes: first, those which cleave substrates that do not permeate the plasma membrane, and second, those enzymes involved with the turnover of envelope polymers during growth. The first group includes nutritionally important enzymes, and some members are well documented; while the second group includes both synthetic and degradative members that have been sparingly studied and, indeed, their existence in some cases is only inferred from experiments with inhibitors.

The methodology for studying enzymes of the cell envelope is generally common to that used with cell-free extracts. However, certain operational advantages accrue because of facile addition and later substraction of substrates or effectors. This is achieved by washing cells (or derivative structures) on the centrifuge or on filters. Conversely, macromolecular probes (enzymes for coupled-assay, labeled antibodies, etc.) are debarred from the periplasmic space by dint of the impermeability of the cell wall, and this may present certain operational disadvantages and problems in interpretation.

This review follows a division according to apparent location of enzymes within the cell envelope. Operational definitions for the component regions of the envelope have been given in Volume I, Chapter 1. It will become evident that the depth of characterization among the individual enzymes varies greatly.

II. PROJECTED BEHAVIOR OF CELL ENVELOPE ENZYMES

Certain criteria can be assembled as hallmarks of this group of enzymes. Other tests help to distinguish fine locations within the envelope, and meeting all or some of these criteria will more or less strengthen the assignment of a particular enzyme to a specific locale. The practical results which speak to these parameters are subject to interpretations that are summarized in succeeding sections. Table 1 is a summary of this type of analysis for the cell wall, periplasmic space, and plasma membrane. Capsulated yeasts may have some enzymes associated with the slime layer, but there is no

Table 1
PROJECTED BEHAVIOR OF CELL ENVELOPE ENZYMES

Criterion	Cell wall	Periplasmic space	Plasma membrane
1. Accessibility to substrate without need of prior perturbation of membranes	+	+	+
2. Modification by reagents excluded by plasma membrane	+	+	+
3. Identity of pH profiles of cell and cell-free extract	+	+	+
4. Release of nonsedimentable activity by protoplasting	+[a]	+	−
5. Release of nonsedimentable activity by cracking.	−	+	−
6. Secretion by protoplasts that are capable of *de novo* enzyme synthesis	+	+	−
7. Cytochemical localization	Yes	Yes	Yes
8. Artificial cross-linking to plasma membrane	−	+	N.A.

Note: (+) = Enzyme in this locale should meet the stated criterion; (−) = should not meet the criterion; and N.A. = not applicable.

[a] May be (−) if enzyme is associated with bud scar residues. See Volume I, Chapter 5, Figures 6 and 7.

information to this effect and in this chapter we can only mention certain salient points about the slime layer in passing.

Criteria 1, 2, and 3 arise from the fact that the cell wall is freely permeated by most low molecular weight substrates, hydrogen ions, and several potential inhibitors or inactivators, that normally do not cross the plasma membrane. (This subject is discussed further in Volume I, Chapter 3.) As well as demonstrating that a candidate enzyme is directly assayable in the intact cell, and that it feels the bathing medium, it becomes equally important to demonstrate that the same conditions do not support assay or modification of cytoplasmic enzymes. If the activity per cell is increased by pretreatments that perturb membranes, then an additional location may be indicated. However, such an increase may be deceptive because the perturbant may also function by inhibiting a secondary event (metabolic or transport) that would otherwise deplete the assayable product in question. This is discussed in more detail under kinetic restraints.

Useful inactivators under criterion 2 include H^+ and heavy metals. The proper functioning of the plasma membrane (see also Volume I, Chapter 4) enables the cytoplasm to be maintained at a fairly constant pH in spite of external changes. Likewise, salts are excluded, especially if the bathing medium is free of metabolizable substrates. However, at extremes of pH the plasma membrane is perturbed and the differential between internal and external pH is abolished. That the integrity of the plasma membrane has not been compromised needs to be demonstrated by the proper controls if criteria 1 to 3 are to be interpreted meaningfully.

Criterion 4 follows from the fact that dissolution of the cell wall by exogenous glucanases will remove the network of the wall, and those enzymes associated with the periplasmic space and the cell wall will now freely diffuse into the bathing medium. Accordingly, the released enzyme will resist sedimentation by high-speed centrifugation. Important controls under this heading include documentation of incidental enzymes in the protoplasting mixture (e.g., snail digestive juice is endowed with many additional enzymes as well as the vital β-glucanases needed for yeast cell wall dissolution) and, in the case of a negative finding, checking the candidate enzyme for susceptibility of inactivation during the protoplasting treatment. Further discussion on protoplast formation is given in Volume II, Chapter 5, but it is worth mentioning here that osmotic vulnerability alone is not sufficient evidence for protoplast formation. That protoplasts devoid of cell wall remnants have indeed been produced needs to be confirmed by electron microscopy.

Release of the enzyme in question during protoplasting does not distinguish between the periplasmic space and the cell wall because the latter is dissolved by the treatment. Criterion 5 provides distinctive evidence in this context because mechanical disruption (See Volume I, Chapter 3 for specific methods) cracks the cell wall in a limited number of locations and large pieces of cell wall are discernible under the light microscope. If the candidate enzyme is associated with the cell wall it will be sedimented by centrifugation, whereas if the enzyme is located in the periplasmic space then but one crack in the wall will be a sufficient opening through which the enzyme can diffuse into the medium. In the latter case there is no need to completely homogenize the wall to generate a nonsedimentable enzyme, and indeed there should be a good correlation between the percentage of cells broken by the treatment and that fraction of the population of candidate enzyme released.

The secretion of enzymes by protoplasts, criterion 6, also follows from the removal of the cell wall, which would restrain those enzymes that normally lodge in the periplasmic space. Incidentally, retention of periplasmic enzymes by regenerating protoplasts should be a good sign for achievement of at least a semblance of functional wall structure (see Volume II, Chapter 6)

At first glance criterion 7, cytochemical localization, would seem to be straightforward. It can be, although several artifactual problems which are general to cytochemical methods are also in play here. As an example one may consider the demonstration of phosphatases by capture of the nascent phosphate as an insoluble and electron-opaque precipitate. Disparate diffusion rates for substrate and trapping agent, or an unfavorable rate of precipitation relative to phosphate ion generation, may lead to artifactual precipitation of phosphate salt some distance from the site of its generation. Additional problems with periplasmic space enzymes relate to their inaccessibility to external enzymes that might otherwise be invoked for a coupled assay (e.g., glucose oxidase and formazan formation as a means to demonstrate sites of glucose generation). This problem of cell wall porosity is not circumvented by working with partly disrupted material because the periplasmic space enzymes will then diffuse into the medium. A study of thin sections embedded in plastic may be feasible for identification of specific compounds, but is unlikely to be worth trying in the case of enzymatic activity. The utility of labeled antibodies in conjunction with thin sections and electron microscopy is feasible but relatively unexplored in yeast.

Criterion 8 is postulated on the rationale of possible cross-linking of candidate enzyme molecules to the nearest neighboring structure. In this context a periplasmic space enzyme might be anchored to the inner aspect of the cell wall as well, although this would depend upon the reactivity of the polymers involved and the choice of cross-linking agent. This approach has been demonstrated recently with periplasmic enzymes from bacteria. Preincubation with diimidoesters makes periplasmic proteins partly inextractable and subsequent separation of envelope components enables some discrimination to be made between the inner and outer membrane as nearest neighbor for a particular enzyme.[3] No published work on yeast cell envelopes along these lines has appeared, but this approach is feasible and will depend on the ingenuity with which potential cross-linking agents are synthesized and applied.

III. ENZYMES OF THE PERIPLASMIC SPACE

Criteria 1 through 6 are met by the β-fructofuranosidase of *Saccharomyces cerevisiae*, while a lesser number of these criteria have been explored for other enzymes that are apparently located in the periplasmic space of various species. Quantitation is very important, i.e., it is essential that bookkeeping be performed on any candidate enzyme through any fractionation scheme. For example, a claim that isolated cell walls are endowed with "high" activity for a particular enzyme is a meaningless statement of localization if the fraction that is still sedimentable is only a few percent of the original total activity. More likely explanations in that particular case are that some adventitious whole cells have escaped the preliminary disruptive treatment or, alternatively, that a small fraction of the monitored enzyme has been artifactually trapped within sedimentable debris. In other cases more than one location within the intact cell may be definitely indicated. Likewise, a similar catalytic activity may be exhibited by more than one protein in a given cell so that some confusion about specific location may result until the two or more proteins are separated and characterized. Documentation along these lines obviously has important implications for precursor to product relationships and for mechanisms of translocation of enzymes destined for the periplasmic space.

A. Detection

The presence of an enzyme in the periplasmic space may be missed for a variety of reasons. Furthermore, quantitative assessments must be made with an awareness of potential problems. The summary that follows is based on reported interferences and means for their rectification and includes also situations that might reasonably be anticipated.

1. Kinetic Restraints

Figure 1 is a diagram of sequential events in the assimilation of a substrate A via intermediates B and C. The regions of the cell envelope are indicated (without regard to scale), and velocities of the individual steps are assigned values of v_1, v_2, ... v_n. Enzyme 1 is assigned to the periplasmic space and carrier-mediated transport of B across the plasma membrane is proposed although that particular mechanism is not essential to the general argument. (A complete discussion of transport processes is to be found in Volume I, Chapter 4.)

If a suspension of cells is incubated with A then compound B may accumulate in the medium according to the scheme in Figure 1, provided that v_2 is larger than v_3. In practice the amount of enzyme 1 is estimated from the maxium velocity for the reaction A → B. The velocity is measured from the rate of disappearance of A or, more commonly, the appearance of B. If v_2 is much greater than v_3 then the rate of appearance of B in the medium will indeed be close to the absolute rate of its generation. But the greater the magnitude of v_3 relative to v_2, then the lower the activity of enzyme 1 will seem by routine monitoring of B in the medium. The latter problem can be rectified by including a specific transport inhibitor in the assay mixture, i.e., v_3 is reduced to zero in the ideal.

Another approach is to add an excess of a coupling-enzyme that will convert B into a nontransported derivative. The coupling-enzyme is unlikely to be able to diffuse through the cell wall, so that its access to B (generated in the periplasmic space) will still

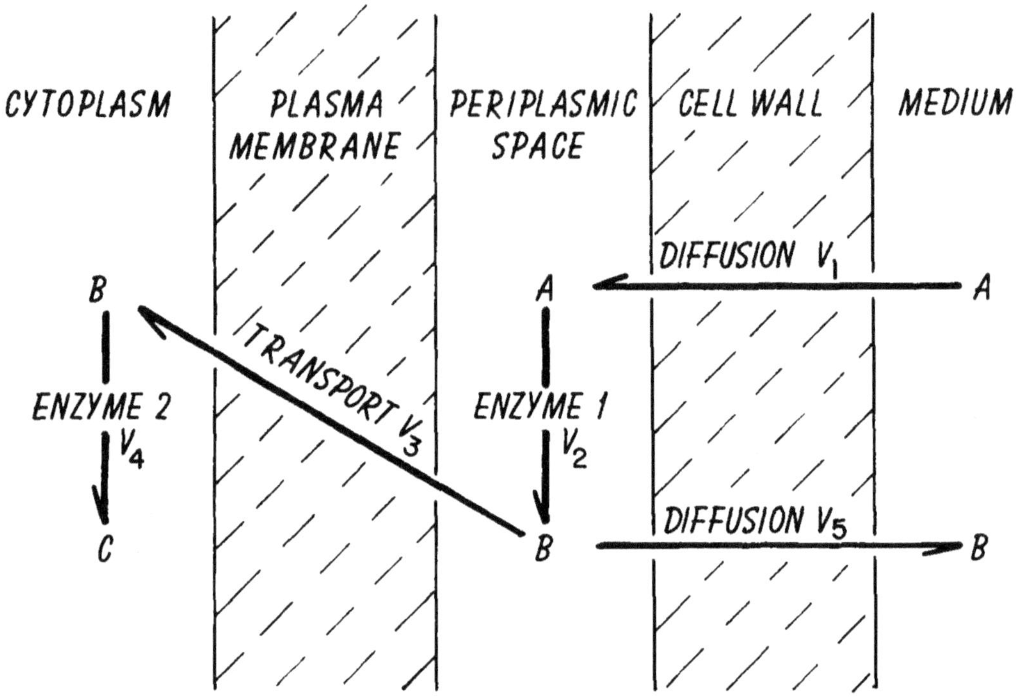

FIGURE 1. Diagrammatic representation of assimilation of a substrate (A) via intermediates (B) and (C). Enzyme 1 is located in the periplasmic space, while Enzyme 2 is in the cytoplasm. Kinetic restraints on the detection of Enzyme 1 in intact cells are discussed in the text in relation to the relative velocities (V) of the enzymatic, transport, and diffusion steps.

depend upon v_5 being substantially greater than v_3; hence this approach is inherently less satisfactory.

If the whole cell suspension is assayed for B (e.g., after stopping the incubation by pH manipulation) it is possible to include cytoplasmic B in the analysis and thus the magnitude of v_4 relative to v_2 will then determine the degree of underestimation of v_2. Addition to the assay mixture of a specific inhibitor of enzyme 2 should result in full values in this circumstance. For that matter, inhibition of some more distant enzyme (in the main pathway for the cytoplasmic metabolism of B) might at least ameliorate the situation because turnover of cosubstrates might be sufficiently compromised to inhibit the pathway. The rational approach is to explore a variety of inhibitors and to discover conditions that maximize activity estimates of the periplasmic enzyme. If other criteria suggest a unique location in the periplasmic space, then the activity of cell-free extracts should also match the values from the optimized cell-suspension assay.

More complicated situations are easily envisaged, e.g., product B might be generated in the periplasmic space and then acted upon by a second extracytoplasmic enzyme. It is not necessary to catalog all of the possibilities, but it is worth considering one specific example in this context. The trisaccharide raffinose (*O*-α-D-galactopyranosyl-(1→6)-*O*-α-D-glucopyranosyl-(1→2)-β-D-fructofuranoside) is fermented by different *Saccharomyces* species to varying extents, as summarized in Table 2. *Sacch. inusitatus* is endowed with an α-galactosidase and a β-fructofuranosidase that are presumably located in the periplasmic space. This would be a case of two enzymes acting on one substrate (raffinose) and although the kinetics are defined for a cell-free system (see, for example, Reference 4) they would be even more complicated for intact cell assay. The actions of the two enzymes on separate molecules of raffinose result in disaccharides,

Table 2
FERMENTATION OF RAFFINOSE BY SELECTED *SACCHAROMYCES* SPECIES

Species	Fraction fermented (moieties)	Residual sugar in medium	Enzymes present, in addition to glc/fru permease		
			β-Fructosidase	α-Galactosidase	Galactozymase
unisporus	0/3	raffinose	−	−	+
cerevisiae	1/3 (fru)	melibiose	+	−	+
oleaceus	1/3 (gal)	sucrose	−	+	+
inusitatus	2/3 (fru, glc)	galactose	+	+	−
uvarum	3/3 (fru, glc, gal)	none	+	+	+

Note: glc = glucose, fru = fructose, gal = galactose. Galactozymase is a collective term for galactose permease, galactokinase, uridyl transferase, and uridine diphosphate-4-epimerase.

sucrose, and melibiose, which subsequently become substrates for the other enzyme in each case. Furthermore, two of the monosaccharides are actively absorbed by the protoplasm whereas galactose is apparently debarred due to the lack of a galactozymase system in this species. The successful assay of the two enzymes in question would obviously require individual substrates, e.g., sucrose and melibiose, in conjunction with an inhibitor of the hexose transport system.

In the case of *Sacch. cerevisiae* the β-fructofuranosidase catalyzed reaction is normally about 300 times more rapid than the intracellular metabolism of the hexoses[5] so that production of reducing sugars from either sucrose or raffinose provides a reasonable assay. On the other hand, one may envisage an advantage in using raffinose for those species that have a similar enzyme complement to that of *Sacch. cerevisiae* but a lower amount of β-fructofuranosidase. This is because one product, melibiose, is not depleted by transport. Such a rationale was employed with mature cells of *Sacch. rouxii.*[6]

2. *Other Causes of Quantitative Error or Nondetection*

A number of conditions influence the assay of an enzyme in the periplasmic space. A rather obvious but essential requirement is that the test substrate be able to penetrate the cell wall. Other aspects such as pH optimum, substrate specificity, and buffer composition are general to all classes of enzymes, are approached by exploring a range of conditions systematically, and need not be labored here. The items that follow deserve special consideration for the present class of enzymes and assume added importance in comparisons among different species for enzyme content and localization. For these reasons the discussion that follows is couched in terms of safeguards that may help to avoid overlooking or underestimating a particular activity.

By the very nature of its location an enzyme in the periplasmic space has direct contact with the medium at all times. The bathing solutions include the growth medium, the washing solution used on harvested cells, and the medium in which cells are stored or further manipulated experimentally. All three liquids may be a source of denaturants, e.g., a transitory pH drop in the growth medium may greatly influence yield of a particular extracytoplasmic enzyme. These conditions usually have less effect on cytoplasmic enzymes. Attention to buffering capacity, pH value, ionic strength, and temperature is important at all stages. Furthermore, the washing medium may elute the enzyme in question as a result of either induced dissociation into subunits or increased porosity of the cell wall. This phenomenon can be approached by assaying all supernatant fractions and by studying the time course of elution, or loss of enzymic activity.

The growth medium may also influence the yield of a particular enzyme through repression of its biosynthesis. Examples include repression of β-fructofuranosidase synthesis by glucose and of acid phosphatase by orthophosphate. Conversely, induction may be necessary. These subjects are discussed in more detail below. The composition of the medium and the quality of the water used in its manufacture are sometimes influential. A recent example from our laboratory involved an acid phosphatase from *Sacch. rouxii* which experienced heavy metal poisoning during a bout with inadequate-quality water that still did not noticeably affect growth.[7] Conversely, the requirement for a metal cofactor must not be overlooked—acid phosphatase from one species of yeast requires Fe^{3+} for full activity.[7] In such cases some consideration should be given to fortifying not only assay mixtures but also growth and washing media.

Recent demonstrations of enzyme inactivation by polypeptide inhibitors and activation by subsequent proteolytic digestion have opened an exciting new field of metabolic control in yeast (see, for example, Reference 8). In terms of covalent modifications the possibilities include protein phosphorylation and dephosphorylation, zymogen forms subject to limited proteolysis, and changes in sulfhydryl status. All types have cytoplasmic examples. To the best of our knowledge no systems of covalent modification have yet been discovered for extracytoplasmic enzymes in yeast.

B. β-Fructofuranosidase

The systematic designation for this enzyme is β-D-fructofuranoside fructohydrolase (EC 3.2.1.26), which is abbreviated to β-fructofuranosidase. Other names that are not recommended by the International Union of Biochemistry, but are used interchangeably in the older literature include sucrase, saccharase, β-fructosidase, and invertase. The latter appelation derives from a change in the sign of optical rotation which attends the complete hydrolysis of sucrose ($M[\alpha]_D^{20°} = 22,770°$) to glucose and fructose ($M[\alpha]_D^{20°} = -7,200°$).

Persoz reported on the inverting action of yeast in 1833, and Berthelot prepared active cell-free extracts and alcohol precipitates of the enzyme as early as 1860. Since then much literature has accumulated on yeast β-fructofuranosidase. Reviews by Neuberg and colleagues[9,10] include interesting historical background and remind us that this enzyme was the subject of landmark studies on enzyme specificity, on the concept and biological importance of pH, and on the kinetics of enzyme-catalyzed reactions. Other reviews that are either devoted to, or contain substantial summaries on yeast β-fructofuranosidase have appeared at intervals over the last 25 years.[11-17] The bulk of the literature is based on *Sacch. cerevisiae* as source material, and the descriptions below pertain to that species unless otherwise stated.

1. Reaction

$$\text{R-}\beta\text{-fructofuranoside} + H_2O \rightarrow \text{fructose} + \text{ROH}$$

The specificity is absolute for the unsubstituted and unchanged β-fructofuranosidic terminal. Thus α-fructofuranosides, fructopyranosides, turnaose and melezitose are not acted upon, whereas methyl-β-D-fructofuranoside, sucrose, raffinose, stachyose, and verbascose are cleaved at the fructosidic linkage[11,12] on the fructose side of the bridge oxygen.[18] Sucrose is the best substrate (lowest K_m, highest V_{max}). Gottschalk[11] postulated that sucrose associates with the active site of β-fructofuranosidase via the glycosidic oxygen, the hydroxyl groups at C_6 and C_3 of the fructofuranosyl moiety, and possibly the hydroxyl group at C_2 of the glucopyranosyl moiety. The latter is obviously not essential because methyl-β-D-fructofuranoside is attacked, albeit less avidly. Substitution of the afructon moiety of sucrose (e.g., the raffinose series) affects kinetic parameters supposedly by dint of steric influence.

The mechanism of the reaction is visualized as an initial formation of α-D-glucopyranose and a fructosyl-enzyme intermediate which then reacts with water to liberate β-D-fructofuranose. Instead of water, any one of the three primary alcohol groups of sucrose may function as an alternative acceptor of the fructosyl group so that in the presence of excess substrate trisaccharides of the kestose type are transient products.[13,19,20] These are eventually cleaved to monosaccharides. The observations that small amounts of two difructosides (C_1 and C_6 substitutions) are also formed during incubation of purified β-fructofuranosidase with sucrose[21] or fructose alone[22] show that fructose can act as both donor and acceptor for the transfer reaction, which is in concert with the above mechanism. A trace of 6-β-fructofuranosylglucose can be isolated from incubations with fructose and glucose but no sucrose is detected.[21,22] Notwithstanding the academic interest in fructosyl-transfer products, there has been no indication that β-fructofuranosidase performs any significant synthetic function in the intact yeast cell.

2. *Assay*

The earlier predilection for the polarimetric method has given way to assays based upon colorimetric detection of the reducing sugar products. My preference is for the 3,5-dinitrosalicylic acid (DNS) reagent of Sumner[23] rather than the copper-reductometric, Somogyi-Nelson procedure.[12] The single-step addition of alkaline DNS reagent serves to terminate the enzymic action, and mixtures require only 5 min in a boiling water bath for color development.[24] The reddish reduction product (3-amino-5-nitrosalicylic acid) has an absorption maximum at 520 nm but we prefer to monitor at 560 nm, where the yellow reagent itself has a low absorption.

Assays of cell suspensions are handled similarly to cell-free extracts. For *Sacch. cerevisiae* the titer of β-fructofuranosidase is so high that suitably diluted suspensions contribute negligible turbidity to the final solution.[25] With less-endowed, and hence more concentrated, suspensions the mixtures need to be cleared by centrifugation after color development.[6,26] In addition, we recommend carrying companion samples through the incubation period (typically 10 to 20 min) at 30°C. These controls receive DNS reagent and substrate (in that order) at term. This enables correction of assay values for any reducing substances derived from other sources in the cells and for any color throw.

One IU of enzyme is that amount which brings about the hydrolysis of 1 μmol substrate (here sucrose) per min at 30°C. Concentrations of enzyme are expressed in units per gram dry weight of cells or per milligram protein. Concentrations of cell suspensions are conveniently estimated by turbidity measurements in conjunction with a standard curve based on dry weight per unit volume.[27] Quantification of cell-free extracts can be based on the original cell suspension or on the protein content. The latter is most conveniently estimated by the Folin-Ciocalteu reagent but the color yield should be established on a Kjeldahl nitrogen basis. For bakers' yeast β-fructofuranosidase the color yield of the purified enzyme is 27% higher than that for bovine serum albumin (the most common standard).[28] Regrettably, corrected protein values are rarely reported.

3. *Location*

A variety of evidence serves to locate the β-fructofuranosidase of *Sacch. cerevisiae* in a region external to the protoplasmic membrane (see, for example, Reference 29 and citations therein). The following findings are convincing evidence that the fine location is in the periplasmic space.

Willstätter and Racke[2] were able to liberate the enzyme by grinding fresh yeast in liquid nitrogen, conditions that would minimize autolysis. Burger et al.[30] used a Hughes' press (see Volume I, Chapter 3) to release the enzyme from fresh or ethyl acetate-treated cells, and Arnold[31] demonstrated corelease of β-fructofuranosidase and acid phosphatase from cells that suffered varying degrees of cracking in the French pressure device (by manipulating the applied pressure). Moreover, the ethyl acetate-treated cell, in which the autolytic enzymes are destroyed, released its β-fructofuranosidase after adding back an endogenous (and partly purified) β-glucanase mixture which was free of mannanases.[32] Hydrolysis of mannosidic bonds would supposedly be required if this glycoprotein (see below) were indeed covalently bonded to the cell wall proper. A claim by Lampen[33] that a specific bacterial mannanase (which splits a mannosidic bond adjacent to a phosphodiester-linked mannose) can release β-fructofuranosidase from live cells is misleading on several counts. The assays involved an additional osmotic shock step; the enzyme preparation still contained some β-glucanase activity; and a further β-glucanase source (snail digestive juice) accelerated and increased the yield of β-fructofuranosidase release.[34]

A model (Figure 2 in Reference 33) which has β-fructofuranosidase covalently bonded to the cell wall of *Sacch. cerevisiae* has been widely quoted, although Lampen[16] subsequently modified his working hypothesis to include the periplasmic space as a second alternative in some species. The former model implies that secondary mechanisms anchor the translocated enzyme to the cell wall matrix. Such a complicated scheme is completely irrelevant to the periplasmic space hypothesis. Furthermore, I feel that the interpretations[33] placed on three seemingly crucial pieces of evidence should be challenged.

First, Sutton and Lampen[35] thought that isolated cell walls had high levels of β-fructofuranosidase; unfortunately, the activity of the cells from which they were prepared was not reported, but we calculate that their wall preparations had only about 3% the activity of commercial bakers' yeast cells (on a dry weight basis). Furthermore, other reports[36,37] indicate very low levels of β-fructofuranosidase in cell wall preparations. Second, Preiss[38] was inclined to locate β-fructofuranosidase in a layer 50 nm removed from the surface and about 50 nm thick. This work was based on inactivation of the enzyme by low-voltage electrons whose energy profiles were theoretically related to penetration of biological material. No attempt was made to estimate the thickness of the cell wall of the test strain, and the fact that *dried* yeast was employed has apparently escaped subsequent reviewers. At best the data only confirm the peripheral location of the enzyme in question. And third, the action of a mannanase[34] on β-fructofuranosidase release has been incorrectly interpreted (see the argument presented above).

The frequent observation that *Sacch. cerevisiae* cells retain all of their β-fructofuranosidase was challenged by Wickerham[39], who was able to demonstrate activity in the medium with some strains. In our hands active cultures of bakers' yeasts have always yielded less than 1% of the total activity in the medium. It would seem more appropriate to describe this phenomenon as a loss of trace amounts to the medium by growing cells rather than substantial secretion. On the other hand, higher percentages were found in the culture fluid from two distillers' strains of *Sacch. cerevisiae* and two strains of *Sacch. uvarum*.[39] We found 20% of the total activity in the medium of a 3-day culture of *Sacch. uvarum* (NRRL Y972), and this is in agreement with the 23% observed previously.[40]

Sacch. rouxii is an exceptional *Saccharomyces* species in that dividing cells produce a physically cryptic β-fructofuranosidase which then becomes expressed (patent) in older cells.[41] We have found that *Sacch. bisporus* var. *mellis* behaves similarly. These species are closely related to *Sacch. bailii* and are regarded as less advanced because they stabilize in nature as haploids. For the last-mentioned species we found two strains with extremely low levels of β-fructofuranosidase and one strain with about half of the activity of *Sacch. rouxii.* Approximately 40% of the *Sacch. bailii* activity was cryptic at 3 days compared with greater than 99% for *Sacch. rouxii* cells.

In contrast to *Sacch. cerevisiae,* the release of β-fructofuranosidase from *Sacch. fragilis* after treatment with sulfhydryl reagents[42,43] and from *Sacch. chevalieri* during osmotic shock[44] is probably due to chemical differences in key structural components of their cell walls, rather than inherent differences in the location of the studied enzyme. In the same general context it is worth mentioning again that we do not know enough about the disposition of components within the cell envelope of *Sacch. cerevisiae,* let alone that of *Sacch. fragilis* (now *Kluyveromyces fragilis*). Unfortunately, some confusion has arisen in the secondary literature because data from different species occasionally have been intermingled with regard to the properties of the cell envelope. Some or all of the above species may be attractive models for further comparative studies on enzymes normally destined for the periplasmic space. In the case of *Sacch. uvarum* a decrease in the molecular size (degree of glycosylation?) of its β-fructofuranosidase, or alternatively an increase in cell wall porosity might be sought to explain the decreased retentivity of those cells relative to *Sacch. cerevisiae.*

An additional β-fructofuranosidase is located in the protoplasm.[45] This fraction is at most 4% of the total activity even in a repression-resistant mutant of *Sacch. cerevisiae.*[46] The internally located enzyme was shown not to be glycosylated, but otherwise to resemble the external variety in molecular weight of the protein moiety and in kinetic parameters.[46] Notwithstanding gross differences in amino acid composition, a biosynthetic relationship between the external and internal β-fructofuranosidases was postulated[46] and received considerable subsequent attention. However, studies involving an inhibitor of glycosylation (tunicamycin) and synchronous cultures negate any precursor to product relationship between the two enzymes.[47] The compulsion to seek a nonglycosylated precursor, and the feeling that such a precursor would be physically cryptic during processing, are founded on higher eukaryotic cell precedents, but for the present the internal species of β-fructofuranosidase which has been described above appears to be independent of the external variety and hence confounds the problem. This leaves the internal enzyme with no known function. Possibly it catalyzes the hydrolysis of the few adventitious molecules of sucrose that penetrate the plasma membrane. Some analogy for the latter occurrence can be borrowed from a mammalian situation: after a high sucrose meal, traces of the disaccharide are found in the urine even though the bulk of the ingested sucrose is first hydrolyzed in the small intestine before entering the portal blood.[48]

4. Chemical Structure and Properties

The association of polysaccharide with yeast β-fructofuranosidase has been of interest ever since the earliest attempts at purification.[9] Willstätter's group had concentrated the enzyme about 1500-fold by 1928, but they were unable to reach a conclusion on its chemical nature.[49] In 1947 Sumner and O'Kane achieved a somewhat higher degree of purification, but more importantly, observed that all of the enzymatic activity was precipitated with concanavalin A.[50] Accordingly, they suggested that this enzyme was a glycoprotein.

Before the invention of suitable mechanical disintegrators the large-scale isolation of the enzyme was almost exclusively by autolysis (see Volume II, Chapter 7), a process suspected to be accompanied by secondary degradation of released proteins and glycoproteins. The use of brewers' or bakers' yeast has also been criticized because of the indefinite genetics. Although the rationale for these two criticisms is quite valid it will be shown below that neither the application of autolysis nor the use of commercial yeasts has really been demonstrated to have a significant effect on the end result.

There is considerably more mannan in the cell envelope than can be accounted for by glycoprotein enzymes (see Volume I, Chapter 5), but the chemical structures of these mannans are quite similar. When broken or autolyzed cells release their β-fructofuranosidase, it is always accompanied by much mannan that is difficult to remove by precipitation and adsorption methods. Probably the best preparation using the older methods was by Fischer and Kohtes.[51] The removal of the remaining noncovalently bonded polysaccharide can be achieved by ion-exchange chromatography on diethylaminoethylcellulose.[28,52-54]

Some examples of purified β-fructofuranosidase are assembled in Table 3. Unfortunately, some published values on enzyme activity employed different units and were measured at some temperature other than 30°C. These values have been converted into international units (IU) by employing appropriate conversion factors and an energy of activation of 11,000 cal·mol^{-1} which is applicable between 0 and 40°C.[57] A specific activity of about 3300 IU (mg protein)$^{-1}$ is indicated for the purified enzyme. Andersen and Jorgensen[55] quoted a polysaccharide content of 13%, but this value is surely down by a factor of 2 judging by the nitrogen content of 11.9%. Apart from the product of Anderson,[52,55] other modern preparations contained 43 to 50% carbohydrate (Table 3). On the other hand, the preparation of Greiling et al.[58], which had only 1600 IU (mg protein)$^{-1}$ and contained 77.4% mannan, must surely be judged impure.

Also of importance is the observation of polydispersity in the carbohydrate content within any sample. I found a continuum between 54% at the front and 39% at the rear of the emergent peak from DEAE-cellulose.[28] The chromatographic behavior was reproducible. When individual subfractions were rechromatographed on a standard column their position reflected their origin in the parent peak.[59] Also, a combined sample of front and rear material gave a bimodal pattern.[59] Fractions which elute from the front of DEAE-cellulose chromatograms exhibit lower anodic mobilities on cellulose acetate electropherograms.[59] This probably indicates that the polysaccharide masks or modifies anionic groups in the protein such that higher contents lead to earlier elution. Berggren[60] reported on differences in carbohydrate content within fractions separated by isoelectric focusing. Polydispersity was rediscovered some 8 years later.[56]

We initially conceived that this polydispersity was due to the autolytic method of enzyme release. However, Neumann and Lampen[54] used mechanical disintegration and found a displacement of the emergent polysaccharide peak with respect to the protein peak from DEAE-cellulose. This would indicate polydispersity even in material prepared in the absence of autolysis. The same laboratory later found increased heterogeneity in the enzyme from two genetically defined strains of *Sacch. cerevisiae* and from a commercial bakers' yeast.[61] The polydispersity thus appears to be a natural phenomenon and variations in carbohydrate content from one preparation to another are probably largely influenced by the "cuts" taken at each fractionation step.

The molecular weight of the protein moiety of β-fructofuranosidase is approximately 135,000.[54] After pronase digestion several glycopeptides were isolated and found to contain mannose and a small percentage of glucosamine residues.[62] Endo-β-*N*-acetylglucosaminidase H from *Streptomyces plicatus* releases most of the carbohydrate as intact oligosaccharide chains. Each chain contains 32 residues of mannose on the average, and two residues of *N*-acetyl-glucosamine, joined to the amide group of asparagine.[56] This type of linkage is discussed in more detail in Volume II, Chapter 2.

Table 3
SOME PURIFIED β-FRUCTOFURANOSIDASE PREPARATIONS

Source[a]	Method of release[b]	Key purification step[c]	Specific activity[d] IU (mg protein)$^{-1}$	Carbohydrate content (%, wt/wt)[e]	Ref.
Bakers' yeast	A	$Al(OH)_3$	1106	72	51
Brewers' yeast	A	DEAE	3574	26	52,55
Brewers' yeast	A	DEAE	3345	47	53
Sacch. cerevisiae (FH4C)	M	DEAE	2850	50	54
Bakers' yeast	A	DEAE	3330	43	28
Brewers' yeast	A	DEAE	3377	48	56

[a] In two cases[52,56] the starting product was a commercial concentrate supposedly isolated from brewers' yeast.
[b] A = autolysis and M = mechanical disruption.
[c] $Al(OH)_3$ = adsorption on $Al(OH)_3$ and elution with lactate buffer (pH 2.9). DEAE = gradient elution from DEAE-cellulose; conditions varied slightly.
[d] In some cases published values had to be converted into international units (IU).
[e] Average values for carbohydrate content, polydispersity is discussed in the text.

According to Neumann and Lampen[62], there is no indication of *O*-glycosidic linkages of the type described for mannan peptides of the cell wall.[63] A claim to the contrary by Greiling et al.[58] can probably be traced to an impure β-fructofuranosidase preparation.

After removal of the bulk of the carbohydrate it is apparently possible to generate two identical 60,000-dalton subunits.[56] This is added evidence to negate an earlier suggestion that internal β-fructofuranosidase might be related to the external variety (in spite of disparate amino acid compositions).[63]

The polysaccharide moiety influences the stability of this glycoprotein towards elevated temperatures. It was found that successive fractions from a DEAE-cellulose chromatogram exhibited increasing sensitivity to inactivation at 65°C; this trend was correlated with a decrease in the polysaccharide to protein ratio.[28] As early as 1919 von Euler and Laurin[65] found that heat inactivation of yeast β-fructofuranosidase (crude extract) exhibited an order of reaction greater than one. Subsequent investigators assumed first-order kinetics for convenience, if not rigor. It is now clear that heat inactivation of unfractionated enzyme is a complex of simultaneous unimolecular processes. It is important to note that heat inactivation of the β-fructofuranosidase in intact cells[31] also departs from first-order kinetics so that the phenomenon in purified enzyme is not a reflection of isolation by autolysis. Discrete fractions, with a narrow range of polysaccharide content, exhibit apparent first-order curves.[28] Chu et al.[66] noted decreased stability in a preparation which was depleted of carbohydrate by treatment with endo-β-*N*-acetylglucosaminidase H. These authors[66] feel that covalently bound carbohydrate plays a role in promoting the refolding of the protein after denaturation with guanidine hydrochloride. Smith and Ballou[67] claimed that the stability towards heat was not compromised by mannan-depletion. However, this is a misnomer since the tests were run at 37°C and at extremes of pH. Moreover, bovine serum albumin was included in all tubes, supposedly to increase stability. At pH values of 1.9 and 8.2 the half-lives were 7.5 and 30 min, respectively, whereas at pH 4.7 there was no detectable inactivation at 37°C.[67] (At pH 5.5, which is the optimum for stability, and 65°C, the half-life is about 7 to 1.4 min and depends on the carbohydrate content.[28]) Inactivation at pH extremes may proceed by a mechanism that is different from that at elevated temperature. Nelson and Kerr[68] suggested that some preparations that are denatured at pH 2.2 may be renaturated upon return to pH 4.5. On the other hand, denaturation as described at 65°C was irreversible.[28]

The pH-activity curve for β-fructofuranosidase shows a broad optimum in the region of 4.5 to 5.5, where the K_m values for sucrose and raffinose are about 26 and 150 m*M*, respectively, and the V_{max} for raffinose is abut 50% that for sucrose.[64] The kinetic parameters for the internal nonglycosylated variety of the enzyme are very similar to the unrelated external form.[64] Ottolenghi[69] made a very thorough study of the influence of pH on K_m and V_{max} for no less than five strains of *Sacch. cerevisiae*, each containing a different nonallelic gene for sucrose assimilation. The 11 kinetic parameters quoted are remarkably constant (see Table 1, Reference 69); for example, the K_s value (which is pH independent) for sucrose varied only from 21.4 m*M* to 27.5 m*M*. This reminds us again of the oft-repeated warning about using defined strains of *Sacch. cerevisiae* as opposed to brewers' or bakers' strains which are polyploid and genetically complex. While it is difficult to argue with the logic, it is curious that so few differences have been brought to light. A similar circumstance pertains to the use of autolysis. Furthermore, Trimble and Maley[56] claim that there is no significant difference between the final product purified from *Sacch. cerevisiae* FH4C by the method of Neumann and Lampen[54] or from a commercial powder.

At least three ionogenic residues participate in the formation of the enzyme-sucrose complex. These have pK values of 8.7, 7.2, and 3.0.[69] The latter is attributed to an essential carboxyl group previously documented by Myrbäck and others (see, for example, Reference 14). Braun postulates that groups with pK values in the vicinity of 3 to 7 participate in the interaction of the enzyme with an epoxide possessing structural analogy to sucrose.[70] However, that assignment[70] of pK values seems to be based on the similarity between pH-inactivation and pH-activity profiles from pH 3 to 7 and an analogy with more detailed analyses of the uninhibited enzyme.[14] The enzyme group with $pK \approx 7$ is assumed to be the imidazole of histidine and is apparently also the locus for inhibition of enzyme activity by Ag^+. Inhibitions by Ag^+ and other metal ions are noncompetitive and reversible.[14] There is no indication of any metal cofactor requirement, so some workers have elected to include 2 m*M* ethylenediaminetetra-acetic acid in assay mixtures to guard against chance contamination by heavy metals in substrate or buffer constituents.[28]

Specific oxidation of 5 to 6 tryptophyl residues per mole of enzyme appears to give complete inactivation, whereas free amino and sulfhydryl groups are not thought to be essential for activity.[71] The insensitivity of β-fructofuranosidase to fluoride at concentrations which inhibit glycolysis[12] is of potential utility in assays of whole cells, especially those species with relatively low activity.[26]

The stability of β-fructofuranosidase, especially in unpurified extracts, is notable. Storage of toluene autolysates of baker's yeast for as long as 19 years results in little loss of activity.[10] Arnold has observed that the ethyl acetate-treated cell, which retains β-fructofuranosidase, is also a remarkably stable source of enzyme activity if suspensions are stored under toluene.[32] Partially purified preparations with specific activities over 2600 IU (mg protein)$^{-1}$ can be dialyzed exhaustively against distilled water, lyophilized, and stored as a stable powder at 4°C.[28] The absence of water improves heat stability[10] and this is of some import in the confectionery industry where β-fructofuranosidase is used in the production of soft-centered chocolates.

The nonglycosylated, internal β-fructofuranosidase of *Sacch. cerevisiae* has been less well studied, but is relatively unstable below pH 6, even at 30°C.[64] That enzyme is also susceptible to inactivation by condensed tannins[16], whereas the glycosylated β-fructofuranosidase is not.[71]

While a great many yeast species have been surveyed for β-fructofuranosidase activity, it is safe to say that only the enzyme from *Sacch. cerevisiae* is well characterized chemically. Because this enzyme is an important marker for the cell envelope, it most likely will be the subject of future comparative studies among different species whose physiological traits suggest that their study may yield some new insight. In this context one should mention that a mutant (FH4C) of *Sacch. cerevisiae* has been used in several laboratories because it carries a single gene (SUC-2) for sucrose assimilation and is novel in a variety of ways. FH4C was selected for relatively high β-fructofuranosidase production, which is not repressible by high glucose in the medium.[54] About 30% of the total enzyme is secreted into the growth medium and the mutant cells flocculate readily.[72] However, the envelope morphology of this mutant is quite bizarre[72] and may contribute to some of the unusual features mentioned. This points up the need for at least an initial cursory survey of ultrastructure on any potentially interesting mutant or species with novel cell envelope chemistry.

C. Acid Phosphatase

The primary role of acid phosphatase (systematic name orthophosphoric monoester phosphohydrolase EC 3.1.3.2) in yeasts is the hydrolysis of external phosphate esters which do not penetrate the plasma membrane.[73] This enzyme has been studied in some depth in a variety of yeast species, although the best documented example is *Sacch. cerevisiae*. Acid phosphatase has one advantage (over β-fructofuranosidase, for example) as a cell envelope marker in that it is subject to cytochemical localization by the long-standing lead capture method.[74]

1. Reaction

$$ROPO_3H^- + H_2O \rightarrow ROH + H_2PO_4^-$$

A single displacement mechanism is involved; the initial products are the alcohol and $(PO_3)^-$, i.e., metaphosphate.[75] The latter reacts with water to generate orthophosphate. In the presence of alcohols, e.g., methanol, *tris* (hydroxymethyl aminomethane), or glucose insignificant degrees of transphosphorylation were demonstrated.[75] This contrasts with much larger degrees of transphosphorylation (60 to 70%) that are demonstrable with acid and alkaline phosphatase from other sources. The latter display a double displacement mechanism involving a phosphorylated enzyme intermediate, whereas the yeast enzyme apparently follows a mechanism analogous to acid hydrolysis.[75]

The purified enzyme is unstable outside the pH 3.5 to 5.5 range so kinetic studies have been restricted to this narrow region. Invariant V_{max} and K_m values with either *p*-nitrophenyl phosphate or β-glycerophosphate as substrate indicate that the ionogenic groups that participate in the enzyme-catalyzed reaction must lie outside the above range.[76] A claim for the participation of a group with pK 3.87 was based on the use of *O*-carboxyphenyl phosphate as substrate[77], but the ionization of the substrate rather than the enzyme was apparently described.[76]

Kinetic studies with *p*-nitrophenyl phosphate and an acid phosphatase from *Schizosaccharomyces pombe* have been reported for an extended range of pH values (2.5 to 6.0). Although a series of pK values (ranging from 2.9 to 5.4) are offered, no assignments to participating ionogenic groups on the protein were hazarded by the authors.[78] Furthermore, the stability[79] of this enzyme is also limited to pH 3.0 to 5.5 at 30°C, so the significance of the exercise is doubtful.

A purified acid phosphatase from *Rhodotorula rubra* is apparently much more stable at 25°C and pH values between 2.6 and 6.0. Even at pH 2.0 only 10% loss in activity per hr was reported.[80] It would be of interest to study the effects of pH on K_m and V_{max} for this enzyme.

While the ubiquity of phosphatases in yeasts is clear, there is still some uncertainty about the plurality of enzymes within a given cell. An alkaline phosphatase[78] is located unambiguously in the cytoplasm and will not be discussed here. Also there is a body of evidence that describes an ATPase associated with the plasma membrane. This enzyme will be discussed below, but should not be confused with the enzyme(s) now under discussion. Whether the extracytoplasmic acid phosphatase activity is due to a single protein possessing a remarkably wide substrate specificity or, alternatively, to possible involvement of several proteins,[82] is still unclear. The situation may depend on the species.

Substrates which are avidly attacked at pH 3.5 to 4.0 by live cells of various *Saccharomyces* species include α- and β-glycerophosphates,[83] hexose phosphate,[83] adenosine tri- and diphosphates,[82] thiamine pyrophosphate,[84] pyrophosphate and linear polyphosphates,[82,85] phenyl phosphate,[82,86,87] *o*-carboxyphenyl phosphate,[77] phenolphthalein diphosphate,[88] and *p*-nitrophenyl phosphate.[31,81] Adenosine monophosphate is attacked very slowly or not at all.[82] A suspension of *Schiz. pombe* attacks most of the phosphate esters mentioned above, but also α-naphthyl phosphate and adenosine monophosphate.[79] Unfortunately, purified preparations[89] of acid phosphatase from *Sacch. cerevisiae* have not been tested against a wide range of substrates but rather the kinetics for β-glycerophosphate and *p*-nitrophenyl phosphate have been studied extensively.[76,89] A purified acid phosphatase from *Schiz. pombe* is better surveyed in this regard; in general the aromatic esters are hydrolyzed at 2 to 3 times the rate of other substrates.[79] A purified acid phosphatase from *Rh. rubra* catalyzed the hydrolysis of a wide range of aromatic phosphates, glycerophosphates, hexose phosphates, and (in common with the enzyme from *Schiz. pombe*[79]) also nucleoside monophosphates.[80] All of the above enzymes are strictly phosphomonoesterases; adenosine triphosphate and tripolyphosphate are attacked stepwise with the liberation of orthophosphate and not pyrophosphate;[82] *bis*-(nitrophenyl) phosphate[79,80] and adenosine 2′, 3′-cyclic phosphoric acid[80] are not attacked. A claim[79] for enzymatic hydrolysis of the last-mentioned compound by the *Schizosaccharomyces* preparation is curious.

2. Assay

Any of the above-mentioned substrates can be employed in conjunction with any one of a variety of methods for measuring orthophosphate as a phosphomolybdate complex. A very sensitive method, the one that we have used most frequently, is due to Chen et al.[90] It does, however, require a 90-min incubation at 37°C for maximum chromogen development. The most frequently cited method by yeast enzymologists is due to Taussky and Shorr[91] and is especially recommended when the less stable phosphate esters are used.

Certainly the most convenient substrate is *p*-nitrophenyl phosphate. A typical assay procedure involves mixing equal volumes of a suitably diluted yeast suspension (containing about 0.5 IU/m*l*) together with 20 m*M* *p*-nitrophenyl phosphate in 200 m*M* sodium acetate buffer (pH 3.8). The substrate and the suspension are equilibrated to 30°C before mixing and incubations continue at this temperature for typically 10 min. At that time 6 vol of 400 m*M* sodium hydroxide are added to stop the enzymatic reaction and to convert the liberated *p*-nitrophenol into nitrophenolate ion which is measured at 425 nm. Although the absorption maximum is 401 nm, we prefer the higher wavelength to avoid the small absorption by the substrate at 401 nm. The incubates are cleared by centrifugation as required and their absorbancies are compared with similarly treated standards of *p*-nitrophenol. Specific activities are expressed on cellular dry weight or protein bases.

Arnold and Lacy[92] showed that *p*-nitrophenol is absorbed by live yeast cells, but this is not a problem because base is added at the termination of incubation and redistribution of nitrophenol to the medium is achieved. Because *p*-nitrophenol is not metabolized by yeasts, the synthetic substrate is recommended for assays involving washed cells or even cells bathed in their growth medium. On the other hand, the uptake of orthophosphate, albeit minimal in the absence of glucose, could sometimes lead to low enzyme values based on phosphate assays.

3. Location

The extracytoplasmic location of acid phosphatase in at least three *Saccharomyces* species was established by work from the laboratories of Rothstein,[82] Suomalainen,[88] Schmidt,[87] Steyn-Parve,[81] Lampen,[94] and Weimberg.[95] These studies accumulated considerable supportive evidence based on accessibility of the enzyme in the intact cell to environmental factors. In addition, it was shown that acid phosphatase was released during protoplasting[93-95] and protoplasts could be made to secrete acid phosphatase into the medium under the right conditions.[96,97]

Tonino and Steyn-Parve[81], working with a bakers' yeast, found that 25% of the total acid phosphatase activity was still sedimentable from cells that had been disintegrated by bombardment with glass beads. Of the total acid phosphatase, 75% is coreleased with β-fructofuranosidase when cells are cracked in the French pressure device.[31] Because the above treatments introduce rents in the wall, these results support the concept that both β-fructofuranosidase and acid phosphatase are restrained by the cell wall rather than covalently bonded to it. Thus the biochemical and biophysical evidence supports a specific location in the periplasmic space.

On the other hand, attempts at cytochemical localization by lead capture methods have, until recently, yielded conflicting results.[98-100] Although these studies sometimes demonstrated lead phosphate deposits in the periplasmic space, they placed more emphasis on the outer surface of the cell wall,[100] suggested that the reaction was restricted to the outer surface,[98] or showed (in frozen sections) a rather nondescript deposition over the cell wall.[99] In contradistinction to the above reports, our own work (see Figure 2) on *Sacch. rouxii* clearly indicated that lead deposits were only formed in the periplasmic space.[7] Cultures of *Sacch. rouxii* (and under some conditions *Sacch. bisporus* var. *mellis*) are atypical of *Saccharomyces* species in that these cells allow a significant fraction of the acid phosphatase to find its way into the growth medium. But the retained acid phosphatase in *Sacch. rouxii* can be assigned to the periplasmic space by biochemical, biophysical, and cytochemical data.[7] We have recently resolved the contradictory cytochemical claims for *Sacch. cerevisiae*.

A long-standing concern with the Gomori reaction for acid phosphatase is the efficiency of capture of the nascent orthophosphate as a lead salt. It is beyond the scope of the present narrative to develop all the arguments and evidence for artifacts. Suffice it to say that if the production of phosphate is rapid, then considerable amounts may diffuse away from the point of generation before precipitating (see, for example, Reference 101). We should also mention that in a biological environment a sol of lead phosphate may form with a net charge and may be delayed from further coagulation by organic substances, charged ions, or pH values below 6.6. A useful biological model that demonstrated most of these points is the uptake of lead by red blood cells.[102]

Our first attempts with *Sacch. cerevisiae* (either bakers' strains or genetically defined strains) gave mixed results. Using a standard cocktail containing 9 m*M* β-glycerophosphate, 3 m*M* lead nitrate, and 50 m*M* acetate buffer (pH 4.25), and 1-hr incubations at

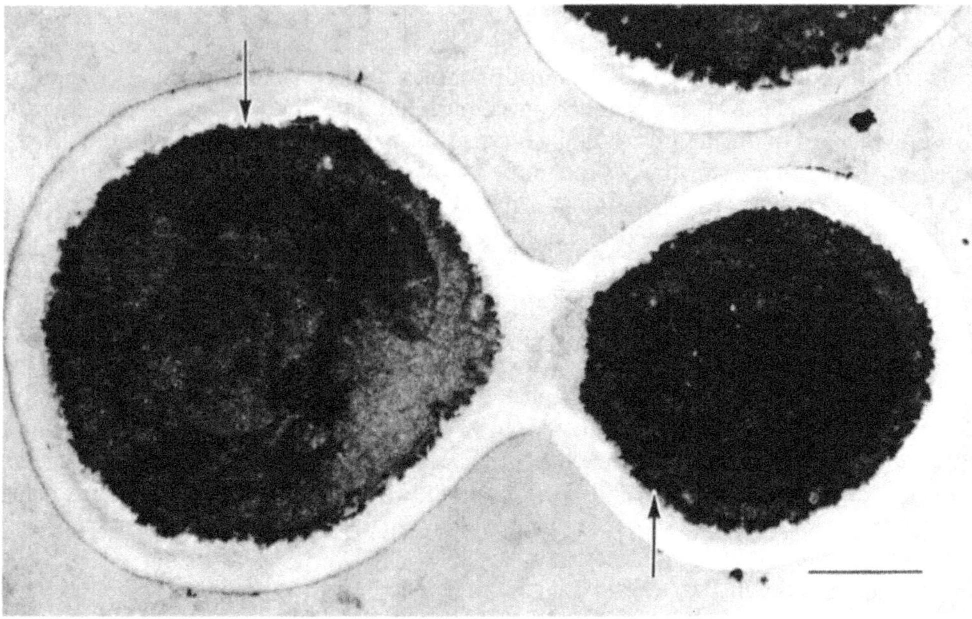

FIGURE 2. Cytochemical localization of acid phosphatase in *Saccharomyces rouxii* by lead capture of orthophosphate generated enzymatically from β-glycerophosphate. Electron opaque deposits of lead phosphate (arrows) are restricted to the periplasmic space. Bar equals 0.5 μm.

37°C, we found some cells with lead phosphate deposits only in the periplasmic space, others with additional deposits in the outer region of the cell wall, and very rarely cells with deposits only on the cell surface. These results circumscribed the above-mentioned reports by others. This was in contrast to our previous results with *Sacch. rouxii* using the same Gomori cocktail. The relatively low activity of the latter species suggested to us that the problem with high-titer *Sacch. cerevisiae* cells might be due to high rates of orthophosphate production that overwhelm the trapping system. Accordingly, we lowered the concentration of β-glycerophosphate in the incubation to 0.25 m*M*, a value predicted to decrease the rate of orthophosphate liberation by a factor of ten. Results with one strain are compared in Figure 3: note that under the low substrate concentration (Figure 3B), lead phosphate is restricted to the periplasmic space. On the other hand, the simplest explanation for the high substrate results is that the outer deposits are artifacts resulting from failure to trap all of the phosphate in the periplasmic space.

At first glance we still had some reservation about this interpretation because we thought that the simple diffusion model would predict deposition of lead phosphate throughout the cell wall. One explanation for the observed gap is that the matrix of the wall will not permit precipitation. Some support for this has come from an experiment in which cells were soaked in lead nitrate and then added to sodium phosphate, i.e., a nonenzymatic analogue to the Gomori reaction. We found that cross sections of these cells also exhibited the two rings of lead phosphate as shown in Figure 3A.

The retention of some acid phosphatase by protoplasts[93,96,97] and the sedimentable nature of some activity with material from disrupted cells[81,83] are most likely due to enzyme embedded in the plasma membrane. This fraction is of considerable interest with respect to mechanisms of translocation and biosynthetic control. These will be discussed later. Membrane may sometimes become attached to broken cell walls and give the impression of activity on the latter.[103] Some overlap with a Mg^{2+} dependent

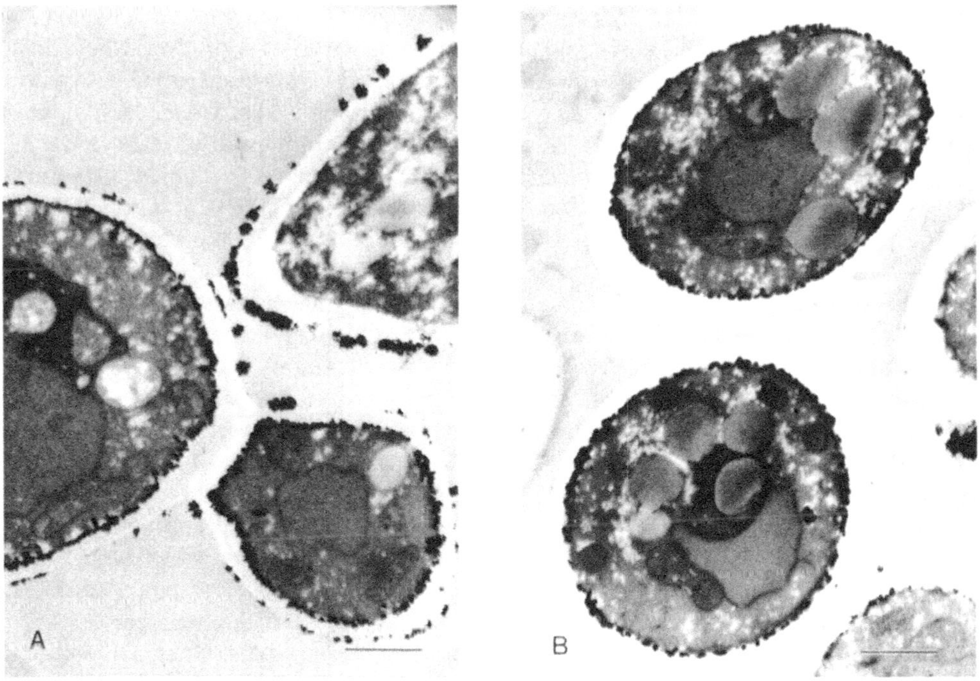

FIGURE 3. Cytochemical localization of acid phosphatase in *Saccharomyces cerevisiae* cells incubated under (A) standard or (B) low concentrations of β-glycerophosphate. Additional deposits of lead phosphate on the cell surface in Figure 3A are deemed to be artifacts. See text for details. Bar equals 0.5 μm in each case.

adenosine triphosphatase (ATPase) may also be involved by assaying this fraction with *p*-nitrophenyl phosphate. There have been some claims for acid phosphatase activity associated with cytoplasmic organelles. Those fractions are beyond the scope of this book, but must represent a very low contribution to the total acid phosphatase activity of the fully derepressed cell. However, for studies on extracts of repressed cells these fractions would need to be taken into account.

The secretion of acid phosphatase by *Sacch. rouxii* cells amounts to about 50% of the total for 3-day cultures.[7] *Sacch. bisporus* var. *mellis*, which is closely related, secretes acid phosphatase into a medium fortified with 0.25 *M* KCl.[85] *Hansenula holsti* cultures have 50% of their acid phosphatase in the medium after 1 day and 100% after 5 days.[104] All three species also secrete extra mannans into the medium; and the tendency to stabilize in the haploid state is also a common property, but the significance, if any, is not known.

Attempts at cytochemical localization of acid phosphatase within *Cryptococcus neoformans* cells were interpreted by Mahvi et al.[105] as indicating activity in the slime layer as well as the cell wall. The published micrographs are reminiscent of earlier attempts with *Saccharomyces* species and should not be interpreted as indicating more than a peripheral location given the above-mentioned restraints on the Gomori reaction. De Nollin et al.[106] reported on *Candida albicans*, but leakage of periplasmic enzymes from their sections (cut from frozen blocks) probably accounts for the cell envelope being devoid of acid phosphatase activity in their hands.

4. Chemical Structure and Properties

All of the purified or partially purified acid phosphatase preparation schemes in the literature have started either with mechanically disrupted cells[89,107,108] or a secreted fraction.[7,85] Autolysis is neither recommended nor practical because the enzyme (from *Sacch. cerevisiae* at least) is destroyed under the same autolytic conditions which do not compromise β-fructofuranosidase.[31] All authors stress the importance of using a culture medium which is low in phosphate in order to achieve a high titer of acid phosphatase in the cellular starting material. Unfortunately, such a medium usually yields cells which are somewhat more refractory to mechanical disintegration (see Volume I, Chapter 3). One recommended purification step takes advantage of the solubility of highly glycosylated proteins in concentrated ammonium sulfate.[85,107] On the other hand, yeast acid phosphatases are invariably precipitated by 50% ethanol or acetone, which may also be used constructively.[80,85,89] Remaining impurities can usually be removed by some form of ion-exchange chromatography,[80,107,108] selective adsorption on calcium phosphate gel,[85] or preparative sucrose-gradient centrifugation.[89]

Properties of the acid phosphatases isolated from five different yeast species are assembled in Table 4. In addition, an acid phosphatase from *Sacch. bisporus* var. *mellis* with pH optimum 5.5 to 6.0 has been purified to a specific activity of 91 IU/mg protein (note, however, that pyrophosphate was used as substrate) and possessed 51% carbohydrate.[85] That species is very close to *Sacch. rouxii.* Two preparations[80,107] gave single and coincident bands on acrylamide gel electrophoretograms which were stained for protein, carbohydrate, or enzyme activity. The same is true for our preparations from *Sacch. rouxii* when examined on cellulose acetate electropherograms.

The carbohydrate content (Table 4) varies among preparations. Heat inactivation of the acid phosphatase in intact cells of *Sacch. cerevisiae* departed from first order kinetics[31]; whether this is due to polydispersity in the polysaccharide moiety (as it is in β-fructofuranosidase[28]) is not known. Broad bands after electrophoresis at some pH values may also be suggesting polydispersity in the enzyme from other species.[80,107] Watorek et al.[80] made a point about the greater degree of heat stability of the enzyme from *Rh. rubra* compared with that from *Schiz. pombe*[107] in the face of similar carbohydrate contents. However, I disagree with their claim[80] that this evidence indicates no contribution from the polysaccharide moiety; supposedly the protein moieties are different.

We have found that the polysaccharide moiety of *Sacch. rouxii* acid phosphatase confers precipitability with concanavalin A. That acid phosphatase preparation and *Sacch. cerevisiae* mannan (see Volume I, Chapters 5 and 6 and Volume II, Chapter 2) gave indistinguishable spectra after reaction with tryptophan-sulfuric acid reagent.[110] After acid hydrolysis, deionization on resins, and paper chromatography of the neutral sugars, only mannose was detected. Mannose was the only sugar detected in hydrolysates of *Sacch. cerevisiae* acid phosphatase.[89] The enzyme from *Rh. rubra* is said to contain mannose, glucosamine (and a trace of fucose?).[80] Acid hydrolysates of the enzyme from *Schiz. pombe* apparently contain about 32% each of mannose and galactose and less than 2% glucosamine.[107] The latter is supposedly involved (as *N*-acetylglucosamine) in the linkage region, but nothing is known about that at present.

A molecular weight of 380,000 ± 30,000 has been estimated from sedimentation measurements on the *Schiz. pombe* enzyme.[107] One rather diffuse band is obtained by electrophoresis in 8 *M* urea and 0.1% sodium dodecyl sulfate; companion data indicate four (supposedly identical) subunits.[107] *C. albicans* acid phosphatase has an apparent molecular weight of 124,000 from ultracentrifugation data.[108]

The K_m values for *p*-nitrophenyl phosphate quoted for *Sacch. cerevisiae* acid phosphatase (see Table 4) beg comment. Boer and co-workers mentioned 1.3 m*M*[89], but subsequently indicated a value of approximately 0.73 m*M* in their Figure 1.[76] The latter study involved substrate concentrations in the range of 0.36 to 2.16 m*M*. Schurr and Yagil[109] quote 0.5 m*M* based on substrate concentrations in the 0.2 to 2.0 m*M* range.

Table 4
PROPERTIES OF SELECTED ACID PHOSPHATASES

Source	pH Optimum	Carbohydrate content (%)	Specific activity[a] (IU/mg protein)	K_m(mM) p-NPP	K_m(mM) β-GP	K_i (mM) $H_2PO_4^-$	Ref.
Saccharomyces cerevisiae	3.8	57	78	1.3	3.1	—	89
	3.7	—	—	0.5	—	0.1	109
	3.5—5.0	—	—	0.7[b]	3.6[b]	—	76
Sacch. rouxii	5.8	82	108	2.0	2.8	2.5	7
Schizosaccharomyces pombe	3.0	67	412	0.2	—	0.5	107
Candida albicans	4.5	88	79	0.3	—	0.1	108
Rhodotorula rubra	4.3	57	365	0.1	—	0.3[b]	80

Note: p-NPP = p-nitrophenyl phosphate and β-GP = β-glycerophosphate.

[a] With p-nitrophenyl phosphate substrate.
[b] Calculated.

Mildner et al.,[111] admittedly working with secreted material from protoplasts rather than a purified enzyme, obtained a biphasic double reciprocal plot relating velocity to *p*-nitrophenyl phosphate concentration. Below 2 mM a K_m of 0.3 mM was obtained, whereas above 2 mM substrate a K_m of approximately 1.2 mM was indicated. The same authors also found two K_m values (0.06 and 0.14 mM) for ATP, which was also hydrolyzed by the same preparation. A brief heat treatment abolished the high K_m for *p*-nitrophenyl phosphate and the low K_m for ATP. Hydrolysis of *p*-nitrophenyl phosphate by the heat-treated preparation was competitively inhibited by ATP, which is in accord with a single nonspecific phosphomonoesterase. The data may indicate that the ATPase activity, also fairly nonspecific (yet clearly possessing a lower K_m for ATP and a higher K_m for *p*-NPP), was destroyed by the brief heat treatment. Mildner et al.[111] still feel that a single enzyme with multiple binding sites is a possibility. So, in this regard, we are not much further ahead on the question of multiplicity of acid phosphatases raised by Rothstein and Meier[82] in 1948.

Yeast acid phosphatases are competitively inhibited by orthophosphate (see Table 4), molybdate, arsenate, and sometimes sulfate, albeit with some variations among enzyme sources. Previously unpublished work on *Sacch. rouxii* acid phosphatase indicates that the K_i values for molybdate, arsenate, and phosphate are 2.5 μM, 4.5 μM, and 2.5 mM, respectively, when *p*-nitrophenyl phosphate is used as substrate. Sodium sulfate (20 mM) had no effect on that enzyme whereas a K_i value for sulfate of 1.8 mM has been reported for the *Schiz. pombe* enzyme.[107]

We found the *Sacch. rouxii* enzyme to be particularly sensitive to certain heavy metal ions.[7] Hg^{2+}, Cu^{2+}, Sn^{2+}, and Fe^{2+} were effective inhibitors at the 1 mM level. The responsiveness towards heavy metals for other enzyme preparations is somewhat varied and the original literature should be consulted. The same is true for fluoride ion which is, in general, not as inhibitory as it is in higher eukaryotic cell counterparts. The alkali metal cations (including K^+ and Na^+) and the alkaline earth metal cations (including Ca^{2+} and Mg^{2+}) have little or no effect on yeast acid phosphatases.

The acid phosphatase of *Sacch. rouxii* is activated by ferric ions.[7] Ferric ions also have an infuence on the secretion of enzyme by growing cells. As far as we are aware, this is the only case of an acid phosphatase from a yeast species which requires a cofactor for full activity. It can be mentioned that certain acid phosphatases from mammalian tissues are activated by Fe^{3+} (see, for example, Reference 112 and references therein) and it will be worth exploring possible similarities in structure and function between these phylogenetically distant glycoproteins. One speculation is that they may be involved in iron transport which is driven by ATP hydrolysis. This brings up the intriguing question of whether the glycoproteins of the periplasmic space have additional functions besides their demonstrated enzymatic activities. The occurrence of levels of hydrolytic activity one or two magnitudes greater than the rates of uptake of the products recommends (to anyone who believes in the economy of nature and the evolutionary process) being alert for other possible roles for these glycoproteins.

D. Other Enzymes of the Periplasmic Space

Additional enzymes seem to be located in the periplasmic space, but the supporting evidence varies and the number of species examined in each case is generally small. Sols and de la Fuente[113] provided evidence that adapted cells of *Sacch. carlsbergensis* (now *uvarum*) split melibiose (6-*O*-α-D-galactopyranosyl-D-glucose) by a hydrolase located in the cell envelope. They showed that the fermentation rate of whole cells was effectively reduced by including hexokinase and ATP, which converted the newly formed glucose into unabsorbed glucose-6-phosphate.

It is worth noting parenthetically that the hexokinase trap was effective in inhibiting sucrose fermentation by *H. anomala* and by *K. fragilis.*[113] Raffinose fermentation by *Sacch. cerevisiae* was similarly inhibited, but sucrose fermentation was not. The latter case was due to the much faster rate of hexose production which overwhelmed the trapping system. As pointed out earlier (Section III.A.1.), this technique yields positive results only when the transport rate is lower than the hydrolytic rate. Because there is no evidence for penetration of the cell wall by the coupling enzyme there is not much advantage to the above technique over any sensitive method for detection of products in the medium.

Evidence for the fine location of melibiase in the periplasmic space comes from the report of Friis and Ottolenghi,[114] who showed that protoplasts from adapted cells of *Sacch. uvarum* lose the ability to ferment melibiose until they resynthesize and secrete melibiase into the medium. The above authors used a glucose oxidase system to measure glucose production from melibiose by cells or protoplasts. Their suggestion for another melibiase fraction in the cytoplasm apparently rests on the increment of activity which attended freeze-thaw pretreatments of cells and protoplasts. Another interpretation is that the transport system which competes with glucose oxidase for liberated glucose led to falsely low values for melibiase activity in the untreated material. Freeze-thaw treatments would inactivate the transport system. Heavy metal inhibitors,[115] which were not investigated in the above work, might give a more definitive indication for this.

According to the assimilation profile (See Table 2) one would suspect that *Sacch. oleaceus* and *Sacch. inusitatus* would also have α-galactosidases in their periplasmic spaces. However, we are not aware of any published data to this point.

Torulopsis glabrata has an extracytoplasmic trehalase which accounts for the high assimilation rate for trehalose by that species.[115] If the transport system for the hexoses is selectively inhibited by heavy metals, then external trehalose is still hydrolyzed and glucose now accumulates in the medium. Furthermore, externally added trehalose leads to a temporary decrease in endogenous concentrations of trehalose within well-fed *T. glabrata* cells.

The latter effect is indicative of extracytoplasmic glucose production because an external glucose signal which triggers internal trehalose mobilization is established, at least for *Sacch. cerevisiae, Sacch. uvarum, C. albicans,* and *T. glabrata* (see References 115 and 116 and citations therein). A cytoplasmic trehalase is clearly indicated for all of the above species. *T. glabrata* apparently has an additional extracytoplasmic trehalase, presumably (but not proven) in the periplasmic space. In this context we should contrast the behavior of *Sacch. cerevisiae* which takes up the disaccharide intact by an acquired transport system.[117]

Kaplan and Tacreiter[118] showed that intact cells of a strain of bakers' yeast can be induced to hydrolyze external β-glucosides. The authors conclude that extracytoplasmic cleavage rather than transport of the glucosides occurs. However, a key experiment showing that label was accumulated by cells fed methyl-β-[^{14}C] glucoside but not with [^{14}C] methyl-β-glucoside is convincing only if it can be shown that methanol inside the cells would be metabolized into nonvolatile components. Otherwise the tracer would have been lost during sample drying, as indeed the authors showed was the case with the cell-free β-glucosidase. I am inclined to believe that we have in this case an inducible transport system for β-glucosides. The other paradoxical and seemingly conflicting data that cellobiose and methyl-β-glucoside are quantitatively used for growth but are not substrates for the internal (cryptic) β-glucosidase[119] can still be explained by lability of the enzyme during the butanol pretreatments of cells. The latter method of exposing internal enzymes apparently is not so damaging to the constitutive aryl-β-glucosidase.[119]

An aminopeptidase activity has been demonstrated for whole cells of *Sacch. cerevisiae* acting on leucine-*p*-nitroanilide.[120] In contradistinction to an additional (and internal) aminopeptidase, the extracytoplasmic fraction apparently does not require Zn^{2+} and is also active on lysine-*p*-nitroanilide. According to Frey and Röhm[120] the latter enzyme, so-called aminopeptidase II, was not secreted by whole cells, but some of the activity was released during protoplasting. Unfortunately, not all of the cells were converted to protoplasts, which is curious because the strain employed (NCYC 366) has been a particularly easy subject for protoplasting by others. The lack of clear-cut accounting on the activity units of aminopeptidase from various fractions suggests that further work will be required before accepting the localizational assignment. If a periplasmic space aminopeptidase is present, this would imply a nutritional function. On the other hand, Marder et al.[121] claim that a variety of leucine-containing and lysine-containing di- and tripeptides are taken up intact by *Sacch. cerevisiae* cells.

IV. ENZYMES OF THE CELL WALL

A. Criteria and Detection

As indicated in Table 1, a cell wall-associated enzyme will meet several of the criteria proposed for those of the periplasmic space. But the distinguishing characteristic is that the enzyme activities under the present heading will be sedimentable after cell wall cracking if the enzyme is firmly associated with the cell wall remnants. Two major operational problems pertain to further characterization.

First, we are concerned about the purity of the cell wall preparation. I have already alluded to misapprehensions due to contamination with unbroken cells. When there are one, or more, orders of magnitude difference in the concentrations of different cell envelope enzymes, then even attaining 99% breakage of cells will not in itself provide a convincing argument for a bound enzyme if individual concentrations in the sediment are not related back to the starting material. A related problem is the plating-out of membrane pieces on the broken cell walls. Monitoring the preparation by electron microscopy is useful, although we are not aware of special procedures for lipids being applied to yeast cell wall preparations. The preservation of phospholipids by osmium tetroxide should be fairly good if they are unsaturated, but the routine dehydration of samples thrugh ethanol and embedding in ordinary epoxy resin is known to cause lipid losses.[122] It would be of interest to apply and to compare one of the polar water-containing embedments (e.g., Reference 123) as a safeguard against artificial purification during sample processing. Another useful monitor is to follow the decrease in specific activity of a second enzyme which, by other criteria, is known not to be associated with the cell wall. All of the published formats should be regarded as enrichment procedures yielding products that approach purity. Several authors are apparently proud of the number of times they have washed their cell wall preparation by sequential steps of centrifugation and resuspension of the pellet in aqueous buffer. Probably a single gradient centrifugation run is more discriminating than ten of the above.

Second, we have the inverse concern about dislodging a cell wall enzyme during the disruptive process. There are no ready answers here, although one approach is to vary the intensity or duration of the disruptive technique and thus to get some feel for the ease of elution. Another potential approach is to attempt cross-linking or polymerization of enzymes within the cell wall matrix prior to disruption. This is unexplored territory.

In terms of detection the major problem is one of enzyme access to added polymeric substrates. The natural substrates for some of the glucanases are insoluble glucans. The inaccessibility of this substrate or of some of the model polymers derived from yeast glucans or other biological sources remains a foremost concern in the demonstration of enzymes associated with the cell wall. For this reason most of the initial documentation has been in terms of autolysis of wall components followed by characterization of the released enzyme which accompanies a certain degree of digestion.

B. Wall Preparations

All cell wall preparations will involve some form of mechanical disruption; this subject is discussed in some detail in Volume I, Chapter 3. The disintegrated cells are then subjected to some regimen aimed at removing unbroken cells, cytoplasmic organelles, and soluble cytosolic components. The format from Nurminen et al.[103] is now briefly summarized to serve as an example.

Yeast cells are disrupted by bombardment of suspensions with glass beads in a Mickle disintegrator (see Volume I, Chapter 3). This and all subsequent preparative steps are performed at 4°C. The beads are removed by filtration through a sintered glass funnel. After centrifugation at 1000 × g for 10 min some of the cytoplasmic organelles are already discarded with the supernatant fraction. Two washes of the pellet by sequential suspension in buffer followed by centrifugation and decantation of the supernate to waste achieve further removal of cytoplasmic components. The pellet still contains unbroken cells, but after centrifugation at 3000 × g for 10 min two distinct bands are visible. The lower band is cream-colored and contains unbroken cells, whereas the upper layer is white and contains mostly pieces of cell wall. The cell wall fraction is elutriated out of the tube by gently stirring with buffer and careful decantation. An alternative procedure is to apply the crude mixture to the top of a continuous gradient of dextran (9 to 21% wt/vol) and to subject this to centrifugation (200 × g for 10 min). The cell walls separate as a band between the pelleted whole cells and an upper layer or layers of membranes and organelles.

C. β-Glucanases

Schiz. pombe cells exhibit hydrolytic activity towards added β-(1 → 3)-glucans, but not β-(1 → 6)-glucans.[124] The molecular weights for the β-(1 → 3) glucan samples employed were approximately 3900 for laminarin and 40,500 for carboxymethylpachyman. In the absence of porosity data for *Schiz. pombe* cell walls one can at least remark on estimates for *Sacch. cerevisiae* walls, which indicate an exclusion threshold of about 4500 daltons (see Volume I, Chapter 3), i.e., about the size of the laminarin but considerably smaller than the CM-pachyman test sample.

No β-glucanase activity was detected in the medium during the time frame in which washed *Schiz. pombe* cells were incubated with CM-pachyman, so that the above evidence points towards hydrolysis of the added polysaccharide by a cell envelope enzyme.[124] Crude homogenates exhibited increased activity, but virtually the total activity was sedimentable, even at low centrifugal forces (approximately 1500 × g). A preparation of broken cell walls, somewhat purified by differential centrifugation, accounted for 55% of the activity in the crude homogenate. This preparation produced oligosaccharides but not free glucose from laminarin and thus an endo-β-glucanase action is indicated. The pH optimum is 5.0. No effects with mercaptoethanol or ethylenediaminetetraacetic acid were found.

The behavior of whole cells contrasted with that for the cell wall preparation in three ways. First, Barras[124] attributed a doubling of activity by the whole cells (but not broken walls) after treatment for 2 hr in 1% (wt/vol) sodium dodecyl sulfate to an opening up of the cell wall structure. This is a reasonable explanation, especially because treatment of cells with 0.5 *M* KCl gave a similar result and there is some precedent for concentrations in this range apparently having an effect on the walls of growing cells of *Sacch. bisporus* var. *mellis*.[85] Second, 0.5 *M* urea completely inactivated cellular activity, but was without effect on the activity of the cell wall preparation. Third, 50 m*M* sodium borate buffer (pH 8.5) decreased the cellular activity only 25%, but destroyed the activity within the cell wall preparation. That solubilization of enzyme accounts for the losses was discounted, because after overnight dialysis of the supernates against 10 m*M* sodium acetate buffer (pH 5.0) Barras[124] could find no significant enzymic activity. Perhaps this is not too unexpected because of the lability subsequently demonstrated for the solubilized enzyme.

Of considerable interest was the finding that a cell wall suspension can liberate oligosaccharides by autolysis.[124] At 30°C rather meager amounts of enzyme were liberated over a 24-hr incubation period in spite of the release of other proteins. The β-glucanase activity of cells in culture was highest in exponential phase and fell significantly during stationary phase. This, together with prior work employing 2-deoxyglucose,[125,126] prompted the suggestion that β-glucanases play a role in cell wall modification during growth.[124]

Megnet[125] found that *Schiz. pombe* cells lyse in late log phase if 2-deoxyglucose is present in the growth medium. Johnson[126] confirmed this and extended the observations to *Sacch. cerevisiae* and *Pichia farinosa*. He introduced the concept that growing cells introduce internal breaks in their cell wall glucan to accommodate expansion, while 2-deoxyglucose interferes with the insertion process of new glucose residues or oligosaccharides. This scheme has been embraced as a working hypothesis by many laboratories, but has not been proven conclusively. In order for normal growth to proceed it is clear that the hydrolytic and synthetic processes on the glucans of the cell wall must be in concert and sophisticated control mechanisms are envisaged. At present it is hard to reconcile the accessibility of wall glucanases (in the intact cell) to externally added laminarin in the absence of indiscriminate autolysis. Broken cell walls autolyze, but supposedly some changes attend disruption. There is no indication yet to indicate that glucan synthesis occurs in the cell envelope in a region distant from the plasma membrane. Perhaps some oligo (1, 3 → 1, 6) transferase activity is involved, but this is purely speculative. This subject is also discussed in Volume I, Chapter 5 and Volume II, Chapter 2.

The effect of 2-deoxyglucose is not as simple as it first appeared. Only growing cells undergo lysis with 2-deoxyglucose and the effect is concentration dependent. Thus when stationary phase cells were inoculated into a fresh, 2% glucose medium fortified with 0.04% 2-deoxyglucose, the viability fell from 100% to 20% after 2 hr.[127] On the other hand, concentrations of 2-deoxyglucose over 0.1% completely inhibited growth, but did not decrease viability within 2 hr.[127] When lysis of cells in the presence of 0.04% 2-deoxyglucose was allowed to proceed to about 40%, further lysis was almost immediately suppressed upon transfer to phosphate buffer. Additions of inhibitors of DNA synthesis or of protein synthesis were effective at the outset in counteracting the effect(s) of 2-deoxyglucose which would otherwise culminate in lysis. These results suggest that *de novo* synthesis of enzymes (or their activators) are required and that turnover of one or more critical proteins is involved. It is possible that the β-glucanase activity found bound to cell walls may only represent a fraction of that participating during normal growth. More recent work (Reference 128 and references therein) has still not defined the mechanism of 2-deoxyglucose action, but points up clear effects on the interconversion of glucose and mannose metabolites and on the orchestration of glycoprotein syntheses.

Fleet and Phaff[129] confirmed the instability of β-glucanase from *Schiz. pombe* cell walls, but in a survey of four additional species found that *Schiz. versatilis* produced eight times as much of the enzyme as did *Schiz. pombe* under the same growth conditions. Accordingly, they elected to study *Schiz. versatilis* cell walls which also have the added advantage of possessing more stable β-glucanases and hence were suitable for isolation and characterization. These workers provided evidence for both an exo- (mol wt 43,000) and an endo-(1 → 3)-glucanase (mol wt 97,000) which were separable by column chromatography on DEAE-cellulose at pH 5.0. They are not glycoproteins. Using laminarin as substrate both exhibit pH optima around 5.0 and the ratio was 7:1 in favor of the exo- variety. This was based on the analysis of enzymes released by autolysis of isolated cell walls.

An interesting finding is that the products of autolysis appear to inhibit further glucanase action.[129] The digestion is taken further towards completion (although complete digestion was never attained in the experiments reported) by incubating cell wall fragments in a dialysis sac. Under the latter conditions 80% of the wall-associated β-glucanase activity was released into the medium in 2 days at 30°C, whereas batchwise incubations released only 45%. Incubation of walls in the presence of laminarin (external substrate) did not increase the release rate; this only indicates that the region of the enzyme which interacts with substrate is different from that which associates with the cell wall structure. After 2 days of self-digestion the walls were not visibly altered under the light microscope whereas after 4 days partial digestion was evident. *Schizosaccharomyces* species have an additional sort of glucan in their cell walls and this polymer is α (1→ 3) linked.[130] The α-glucan apparently contributes to cell wall integrity; isolated walls are devoid of α-glucanase activity.[124,129]

The problem of cell wall purity was addressed by Fleet and Phaff,[129] who were aware that glucanases could have become attached during the isolation process. The weight of evidence[124,129] supports a firm association of some glucanases with the cell wall, although the mention[129] of β-fructofuranosidase, acid phosphatase, and proteolytic activities associated with purified cell wall preparations is disturbing. In this connection it is significant that both endo- and exo-β-glucanases are also located in cytoplasmic vesicles.[131] My own work[32] (and see Volume II, Chapter 7) on the autolytic enzymes of bakers' yeast showed the presence of active β-glucanases in cell-free extracts from live cells pretreated with 0.1 *N* HCl for 1 hr. This indicates at least one additional source, i.e., the cytoplasm, for autolytic glucanases in that species. There is a plethora of endogenous β-glucanases in yeast species generally and in some cases they are also secreted into the medium.[132-134] One approach to localization is the generation of protoplasts, but unfortunately this invariably involves adding a concentrated external glucanase and makes analysis of the liberated products difficult to interpret. Washed protoplasts of *Sacch. cerevisiae* will secrete β-glucanases into the medium and because the process is inhibitable by cycloheximide, it is felt that these represent *de novo* syntheses rather than relocations of internal enzymes or elution of β-glucanases originally from the protoplasting mixture.[133] Villa et al.[134] provided evidence for three β-glucanases in *P. polymorpha* and, using the technique of relative accessibility to acid inactivation,[32] concluded that the bulk of the activity was extracytoplasmic. Because their cell wall preparations were devoid of activity, they[134] assigned a periplasmic space location. Contrast this with the cell wall location claimed for the β-glucanase of *P. pastoris.*[129]

Beginning with the report by Brock[135], there have been several literature references to the participation of β-glucanases in the conjugation process between mating haploid cells. Here a partial dissolution of adjacent walls is involved and some leakage of enzyme to the medium seems to attend. For the vegetative stage some preliminary evidence has been presented[129] to support the idea that isolated walls from *K. fragilis, H. anomala, Sacch. cerevisiae, Sacch. rosei, P. pastoris,* and *C. utilis* all contain significant amounts of β-glucanases.

D. Other Cell Wall Enzymes

It would seem almost inescapable that further examples will be discovered, but no other candidates fulfill enough of the criteria at present to be so assigned specifically to the cell wall. It is clear that all enzymes detected in the culture medium must leave the cell via the wall. The time that they spend in association with the cell wall may depend on the nature of the medium and the age of the cell.

Ahearn et al.[136] surveyed a number of yeast species for proteolytic activity (against casein or gelatin) and found several positive indications, especially among *Kluyveromyces* species, for which the cell-free filtrates of the media were negative. Chen and Miller[137] found an increase in a protease (against denaturated hemoglobin at pH 3.7) activity of whole cells of *Sacch. cerevisiae* preceding sporulation. On the basis of limited penetration of the cell wall by the substrate and of the absence of activity in the medium, these authors are inclined to attribute the protease activity to the periphery of the cell wall. The pH optimum and physiological role, if any, is a puzzle in that the pH of the sporulation medium started at 7.5 and rose to about 8.7 during the incubation. Nevertheless, these reports[136,137] may suggest useful biological material for future studies aimed at exploring possible control mechanisms involving limited proteolysis of cell envelope enzymes.

At the risk of belaboring the problems of cell wall preparation, I am convinced that further advances in this area will only proceed after new or modified methods are developed. It would be of interest, for example, to equilibrate cells with a bathing solution of high osmotic pressure prior to disruption. The aim here would be to decrease the chance of membrane association by having the protoplasm pull away from the cell wall prior to disruption. Unfortunately, this gives a more refractory cell (see Volume I, Chapter 3), at least in response to shearing stresses. Another approach that seems to be worth trying is to induce some chemical modifications (cross-linking, polymerizations) in the cell walls prior to their isolation. The putative importance of endogenous wall-modifying enzymes to growth mechanisms recommends further study in this obviously difficult area.

V. ENZYMES OF THE PLASMA MEMBRANE

The plasma membrane performs a multitude of functions affecting the flux of small molecules between the protoplasm and its environment. These permease systems involve specific proteins which enter into complexes with the transported substances and are discussed in Volume I, Chapter 4. In addition, the plasma membrane has certain hydrolytic and synthetic activities which seem to be uniquely associated with this region of the cell envelope.

A. Criteria and Detection

An enzyme may be embedded in the plasma membrane such that its active center is directed towards the outside (i.e., periplasmic space) or the inside (i.e., cytoplasm). In the first case the enzyme may be assayable in the intact cell. Kinetic and other restraints based on relative permeability of potential substrates, will influence the quantitative result in much the same way as outlined for enzymes in the periplasmic space. Moreover, by cytochemical test one would anticipate reaction product on the outside of the plasma membrane and in the periplasmic space, so that a fine distinction would probably not be possible by this criterion alone. In the second case, some form of membrane disruption is a prerequisite for quantitative assay. Here the paramount concern is that the integrity of the broken membrane is not compromised to an extent that precludes native enzyme activity.

Two forms of latency are theoretically possible in the present context. First, an enzyme (or at least its active site) may be temporally disposed in the hydrophobic region of the phospholipid membrane and thus inaccessible to hydrophilic substrates even in broken membrane preparations. And second, an inactive zymogen form may require limited proteolysis before catalytic activity is expressed. (A different mechanism, but with the same end, would be the association of enzyme with an inhibitor, and the latter would need to be removed for activation.) These mechanisms are mentioned here because there is already one documented case (chitin synthetase), and others are anticipated because control mechanisms on membrane enzymes are postulated to be important in the growth of cells and in their adjustments to environmental changes.

There is also considerable current interest in the association of particles and vesicles with the plasma membrane and the possible roles they play in the translocation of polymers destined for more external regions of the cell envelope or the medium. This subject is reserved for Section VIII. B.

The locational assignment of enzymes to the plasma membrane is very much strengthened by the relative success in removal of contaminating membrane fragments. The latter arise from disruption of cytoplasmic organelles and are sometimes difficult to separate by straightforward centrifugation methods. Accordingly, all plasma membrane preparations reported in the literature represent approaches to purity whose relative achievement must be judged from the content of other marker enzymes and their appearance in electron micrographs.

B. Membrane Preparations

There are two basic approaches. One is to generate protoplasts, lyse them, and then separate the membranes.[138] The second approach, which originated in Suomalainen's laboratory, is to mechanically disrupt cells, separate walls and adhering plasma membranes, and then digest the cell walls with added glucanases.[84,139] Dube et al.[140] have examined the technique of Nurminen et al.[84] primarily from the morphological point of view. They found no evidence of adhesion of membranes to the inner aspect of wall fragments, but did find entrapment of various vesicular fractions. Digestion of wall fragments was not uniform and, on the basis of electrophoresis patterns on the extracted proteins, contamination by mitochondrial membranes was present even in the purified preparations. The protoplast route is in vogue presently, although the above alternative (or some variation therein) should be kept in mind. Two fundamental differences are worth noting. Enzymes associated with the plasma membrane that are disposed towards the interior in the intact cell will not see the digestion mixture used in protoplast generation, but they will in the cell wall preparation. On the other hand, by working with large pieces of broken cell wall one avoids subjecting membranes to possible artifacts resulting from osmotically-fortified media.

The lysis of protoplasts can be achieved mechanically,[141] by dilution of the osmoticum,[138] or by metabolic lysis.[142] Most workers in this field have included 1 to 10 m*M* Mg^{2+} in the lysing fluid. Scarborough[143] introduced the idea of reinforcing fungal plasma membranes with concanavalin A and this has been applied to yeast.[144] Abrams[145] showed that bacterial protoplasts tend to lyse in the presence of glucose, and Indge[146] was the first to exploit this phenomenon with yeast protoplasts. This technique has the advantage that by careful manipulation of the ionic environment during lysis one may protect vacuoles from breaking and hence avoid exposing the plasma membrane to some proteolytic enzyme.[142]

The bulk of the yeast work in the literature pertaining to enzymology of the plasma membrane is on *Sacch. cerevisiae* and *Sacch. uvarum*. However, *C. albicans*[147,148] and *C. tropicalis*[149] plasma membranes can be isolated by similar procedures. Duran et al.[144] find that concanavalin A flocculates protoplasts of *Sacch. cerevisiae*. When the flocs are ruptured (by decreasing the osmotic pressure and blending) the plasma membrane is obtained as essentially a single piece per cell. Without the reinforcement by concanavalin A only small membrane fragments are obtained. In either case membrane preparations are washed by sedimentation in the centrifuge at approximately 100,000 × g for 30 min. Purification is obtained by centrifugation in a gradient constructed with meglumine diatrizoate,[144] sucrose[141,147] or Ficoll®,[142] or by differential centrifugation.[84,138]

The identification of the gradient fraction which is enriched in plasma membrane is made by electron microscopy and also by measuring the distribution of a label which is applied to protoplasts before lysis. Suitable labels are radioactive iodine,[144,147,151] radioactive[144] or fluorescein-conjugated[147] concanavalin A, and radioactive *N*-ethylmaleimide.[150]

C. Chitin Synthetase

A plasma membrane preparation from *Sacch. uvarum*,[152] and a solubilized enzyme from the plasma membrane of *Sacch. cerevisiae*,[153] catalyze the incorporation of *N*-acetylglucosamine from UDP-*N*-acetylglucosamine into a polymer indistinguishable from normal chitin. This chitin synthetase (EC 2.4.1.16) has a K_m for UDP-*N*-acetylglucosamine of about 0.8 m*M*, requires 10 m*M* Mg^{2+} for maximal activity, and prefers a pH in the region of 6.2. Stimulation of particulate activity by 10 m*M* *N*-acetylglucosamine is probably not of physiological significance. Furthermore, the solubilized enzyme is not very much affected by this sugar. The other product of the synthetase reaction, UDP, exerts 62% inhibition at a concentration of 0.5 m*M*, but once again this is probably not of physiological significance. The antibiotic Polyoxin A (which is a structural analog of UDP-*N*-acetylglucosamine) is a very strong inhibitor ($K_i = 0.5\ \mu M$) of yeast chitin synthetase.

On the basis of susceptibility to glutaraldehyde inactivation in isolated membranes as opposed to resistance in the intact protoplasts, Duran et al.[144] concluded that chitin synthetase is disposed on the plasma membrane towards the interior of the cell. This result actually means that the part of the enzyme which needs to see the substrate

N-acetylglucosamine is directed inwards; the product, chitin, would almost certainly need to be released outwards. There is no indication at present for any endogenous acceptor being required[153], although the critical experiment has not been done.

Most of the chitin synthetase in the intact cell or isolated membranes of yeast exists in an inactive zymogen form.[154] Limited proteolysis expresses the full activity.[152] Trypsin and other proteolytic enzymes are effective, but of the several proteinases in yeast only proteinase B seems to be active.[142] The proteinases are normally restricted to the vacuole, and some vesicular transport system is postulated to be involved in the activation of chitin synthetase. A heat-stable inhibitor of proteinase B exists in the cytosol.[155] This inhibitor is now known to be identical to a chitin synthetase inhibitor which prohibits activation, but has no effect on trypsin-activated chitin synthetase. Hasilik[156] showed that incubation of chitin synthetase with higher levels of proteinase B eventually leads to inactivation, apparently through more extensive proteolysis. The discovery that mutants of *Sacch. cerevisiae* that lack proteinase B activity can still successfully bud and synthesize chitin throws considerable doubt on the earlier indication that proteinase B is the natural activator of chitin synthetase.[157,158]

The synthesis of chitin in synchronously dividing cells is timed to the appearance of buds and occurs stepwise as opposed to the continuous incorporation of dry matter.[159] Stationary phase cells are devoid of active chitin synthetase, but have some enzyme in the zymogen form.[156] Because of the putative function of chitin synthetase in yeast morphogenesis, the localization of the enzyme on the plasma membrane is a vital part of this working hypothesis. Schekman and Brawley[160] found that intracellular vesicles (from haploids that were stimulated by mating factor) also possessed some chitin synthetase activity. Nevertheless, chitin synthetase is presently the most useful enzyme marker for the plasma membrane. Several other fungal species have been examined for chitin synthetase; these are beyond the scope of the present narrative, but the reader is referred to Ruiz-Herrera and Bartnicki-Garcia[161] for work on the yeast-like cells of *Mucor rouxii*.

D. Mg^{2+}-Adenosine Triphosphatase

A Mg^{2+}-dependent ATPase has for some time been suggested as a marker for the yeast plasma membrane. Until recently there was still considerable controversy as to the locational assignment; the problem has been largely one of membrane fractionation and the confidence one could attach to the various so-called purified preparations. Nurminen et al.[84] undoubtedly demonstrated such an activity, but only 18% of the total was oligomycin resistant. Moreover, criticism[140] of the isolation technique[84] threw some weight on the possibility that membranes from mitochondria or other sources were contributing significantly. However, recent work on *Sacch. cerevisiae* and *Schiz. pombe* membranes has confirmed the existence of a plasma membrane ATPase with pH optimum in the 5.5 to 6.0 range and dependence on Mg^{2+}.

Willsky[141] has prepared a membrane fraction from lysed protoplasts of *Sacch. cerevisiae* which, on the basis of apparent density (1.22 g/ml) in a sucrose gradient, she equates to broken plasma membranes. An ATPase activity was demonstrated and had the following properties: K_m for ATP = 1.7 m*M*, pH optimum around 5.5, and Mg^{2+} requirement of 5 m*M* (i.e., about equimolar with ATP). At the pH optimum 60% of the ATPase activity in the partially purified membrane fraction was resistant to oligomycin (133 μg/mg protein) and mention has been made of decreased sensitivity to Dio-9 and Efapeptin (known inhibitors[162] of mitochondrial ATPases). On the bases of no influence (Ca^{2+}) or slight inhibition (Na^+ or K^+) by metal ions, resistance to Strophanthin (a specific inhibitor of Na^+/K^+ ATPases), and pH optimum, it has been concluded[141] that this yeast enzyme is unlike that in mammalian metal-ion pumps. However, on the bases of vanadate inhibition (K_i – 11 μM) and some preliminary data on the formation of three transient phosphoprotein intermediates, the mechanism of the yeast enzyme may be more akin to that of mammalian Na^+/K^+ ATPases than to bacterial or mitochondrial Mg^{2+}-ATPases. According to Willsky[162a], sorbitol inhibits Mg^{2+}-ATPase from the yeast plasma membrane. This may help to resolve the negative results voiced by some workers who prefer sorbitol (or mannitol) as osmoticum during protoplast generation. Schibeci et al.[150] feel that Mg^{2+}-stimulated ATPase is easily released into the soluble fraction and hence unsuitable as a marker for the plasma membrane.

Goffeau and co-workers[163,164] disrupted cells of *Schiz. pombe* in a Braun disintegrator and then, through a sophisticated scheme involving six rate centrifugation steps and a pH manipulation, obtained a fraction rich in plasma membranes. The final approach to purification involved an equilibrium centrifugation run on a discontinuous sucrose gradient. The band of particles with density 1.20 g/m*l* contained a Mg^{2+}-dependent ATPase said to be enriched about 90-fold. A solution containing lysolecithin (2 mg/m*l*), ATP (1 m*M*), and EDTA (1 m*M*) extracted the ATPase; optimally at pH 7.5. The solubilized enzyme was purified by centrifugation through a linear sucrose gradient. The active fraction gave a single protein band (mol wt approximately 100,000) on SDS-polyacrylamide electrophoresis together with a different and unidentified polysaccharide band. The purified enzyme is stimulated by the presence of lysolecithin and inhibited by *N,N'*-dicyclohexyl carbodiimide and Dio-9. In this last respect the ATPase from *Schiz. pombe* mitochondria is also sensitive, which is general for other mitochondrial enzymes, but contrasts with the behavior of the ATPase from *Sacch. cerevisiae* plasma membranes.[141]

E. Phospholipase

Nurminen and Suomalainen[165] disrupted bakers' yeast cells in a Mickle disintegrator and subjected the fragments to partial digestion with snail gut enzymes. (The latter were freed of lipolytic activity by gel permeation chromatography.) By differential centrifugation they were able to obtain a fraction which was at least enriched in plasma membrane pieces (see, however, Reference 140 for reservations). This preparation contained Mg^{2+}-dependent adenosine triphosphatase and phospholipase activities. Lipase activity was weak.

The facile elution of the phospholipase activity by dilute tris buffer[165] was reminiscent of the elution of phospholipase A and acyltransferase from intact yeast. Thus, Kokke et al.[166] found that lysolecithin, which has strong detergent action on a variety of cell membranes, had little effect on intact bakers' yeast because it is quickly converted into fatty acid and glycerylphosphoryl choline. The responsible enzymes are rather easily eluted from intact cells by distilled water and this strengthens the locational assignment by Nurminen and Suomalainen.[165] However, some caution must be exhibited because their preparations contained small amounts of β-fructofuranosidase and the phospholipase may have originated from another source.

F. Glucan Synthetase

Studies on the synthesis of β-glucans by permeabilized cells,[167] disrupted cells,[168] cell wall fractions,[169] and membrane fractions,[168,169] of *Sacch. cerevisiae* have shown incorporation of labeled glucose from UDP-glucose or GDP-glucose into β (1 – 3) linked polymers, or in one case,[168] a polymer having β (1 – 6) linkages as well. Because of the low activities generally encountered and the mixed membrane fractions employed, little could be said about either the number of enzymes involved or their localization.

Recent communications[170,171] from Cabib's laboratory report that fairly high rates of incorporation can be achieved when crude membrane preparations are fortified with glycerol, bovine serum albumin, and ATP or GTP. Although recoveries were low, the activity seems to be restricted to the plasma membrane (prepared in similar fashion to that used in chitin synthetase studies).

Only UDP-glucose was needed as glucose donor and good evidence was assembled to indicate that no polysaccharide primer was required[170] The synthetase required no metals and there was no indication of lipid intermediates.[170] The pH optimum is in the vicinity of 8.0 and the K_m for UDP-glucose is 1.7 m*M*. On the basis of short term glutaraldehyde accessibility, the authors concluded that the site of the protein which sees UDP-glucose is disposed towards the interior of the cell. Glucan synthetase apparently does not rest in a zymogen form[170], but is influenced by nucleoside phosphates in a seemingly complex manner.[171] Some evidence was presented to imply that another compound mediates these modulating effects and that this compound is inactive when dephosphorylated.[171] It is clear from the above studies that glucan synthetase is influenced by its location in a membrane, that disruption has profound effects, and that the lipid environment may play a role, which is obviously experimentally difficult to probe.

G. Other Possibilities for the Plasma Membrane

It has been claimed[151] that 20% of the mannosyl transferase (incorporation of mannose from GDP-mannose into α-mannan) in crude membrane preparations from *Sacch. cerevisiae* resides in the plasma membrane. However, the finding that 80% of the activity sedimented in other membrane fractions seems to argue against putting too much stock in the plasma membrane component at this time. Likewise, evidence of glycogen synthesis from UDP-glucose by the same membrane fragments needs further substantiation. However, one can mention in passing the proposal by Gunja-Smith[172,173] and colleagues that some glycogen resides in the periplasmic space. That claim has not been further substantiated or disproved, to the best of my knowledge.

Wheeler et al.[174] studied the distribution of adenyl cyclase within particulate subfractions of lysed protoplasts. They report very little activity for the fraction deemed to be richest in plasma membrane. Londesborough and Nurminen[175] found a Mn^{2+} dependent adenyl cyclase in whole homogenates of *Sacch. cerevisiae* and attributed a considerable part to the plasma membrane. That this enzyme is membrane-bound seems clear,[174,175] but it is apparently widespread among various disrupted membrane fragments.[176]

VI. ENZYMES YET TO BE DIFFERENTIATED WITHIN CELL ENVELOPES

Some strains of *Sacch. cerevisiae* have an extracytoplasmic asparaginase which is induced under conditions of nitrogen starvation.[177] Cells so endowed attack both L- and D-asparagine and release ammonia to the medium in short term incubations. Prolonged incubations indicate ammonia absorption by cells. All strains of *Sacch. cerevisiae* apparently produce a constitutive cytoplasmic asparaginase which catalyzes the hydrolysis of L-asparagine to L-aspartic acid and ammonia. The inducible form is a glycoprotein with polydispersity in the mannan content.[178] That this enzyme is subject to nitrogen catabolite repression[179] and is genetically distinct from the internal L-asparaginase[178] may recommend it as a subject for future studies. This enzyme would seem to be a good candidate for localization in the periplasmic space, but unfortunately no studies along these lines have been presented to date.

A fraction of the catalase activity of *Sacch. cerevisiae* can be detected *in situ*. This fraction is sensitive to inactivation by heavy metal ions that do not easily penetrate the plasma membrane (see Reference 180 and citations therein). However, fine location has not yet been attempted. The unusual behavior of the cryptic fraction of catalase under heat-denaturing conditions[181] is also interesting and might be amenable to study in protoplasts.

Singh and Datta[182] find that protoplasts of *C. albicans* can be induced to synthesize a kinase, a deacetylase, and a deaminase whose sequential actions convert *N*-acetylglucosamine into fructose-6-phosphate. That part of the newly synthesized kinase and deaminase are secreted would suggest a partial location within the periplasmic space of induced cells. A possible involvement in the further metabolism of chitin degradation is purely speculative at this juncture.

VII. ENZYMES SECRETED INTO THE MEDIUM

Various enzyme activities have been associated with cell-free filtrates of growth media. In some cases those are thought to be derived only from autolyzing cells and will not be further discussed in this chapter. Other examples are clearly derived from viable cells and need to be mentioned here if for no other reason than that they pass through the cell wall.

Snyder and Phaff[183] performed several tests on the β-fructosidase of *K. fragilis*. They have remarked on the significant fraction which is secreted into growth medium, but found no difference between the secreted and retained fractions. Furthermore, the same enzyme apparently acts on inulin, raffinose, and sucrose. Although Snyder and Phaff were of the opinion that they were dealing with an inulinase, which is distinguishable from the β-fructofuranosidase (invertase) of *Sacch. cerevisiae* in its relative affinity for inulin; this more likely points up a small difference between the enzymes of the two species as well as the apparent difference in cell wall porosity.

Bakers' yeast autolysates completely hydrolyze inulin, but intact cells do not ferment high molecular weight inulins. Weidenhagen was of the opinion (see Reference 183) that high molecular weight inulins are unable to penetrate the cell wall. This would be in sympathy with contemporary thinking about cell wall structure (see Volume I, Chapter 3).

Different strains of *Sacch. diastaticus* adapt more or less quickly to an amylopectin-containing medium and secrete glucamylase within a few hours or, in extreme cases, after several days.[184,185] Although the phenomenon is of considerable interest to the brewing industry in the production of so-called light beers, very little definitive enzymology has been published.

Endomycopsis fibuligera, which exhibits abundant true mycelium as well as budding yeast cells, produces an extracellular starch degrading system.[186] The hydrolytic activity on soluble starch has been monitored in different growth media, but the enzymes remain to be characterized. With the heightened interest in production of alcoholic fuel from starch sources, it is anticipated that some spin-off on the cell biology of these enzymes will accrue.

Villa and associates have isolated and purified two β-glucanases which are secreted by *C. utilis*; one is an endo-1,3-β-glucanase[187] and the other has an exo-1,3-β-glucanase action pattern.[188] On the other hand *Candida* species do not secrete α-(1 → 3), β-(1 → 3), or β-(1 → 6) glucanases into culture medium.[189] Wall preparations from *Cr. albidus* possess all three of the above activities and apparently small amounts of β-(1 → 3)-glucanase are also secreted by growing cells.[189] Abd-El-Al and Phaff[190] demonstrated endo-β-(1 → 3)-glucanase activity in the culture fluid of *Hanseniaspora valbyensis* and *H'spora uvarum*. In these cases the secreted activity is not stimulated by the addition of glucans to the growth medium. Exo-β-glucanase activity has been detected in the culture fluid from aerobically grown *K. fragilis* and *H. anomala*.[191] In most of these examples it is not clear how much of the respective β-glucanases are released incidentally to the cell wall modification that accompanies growth. Supposedly the inherent variations in porosity of different cell walls also have an influence on the amounts secreted.

Ahearn et al.[192] have carried out a preliminary survey on yeast and yeast-like fungi for extracellular proteinases. On the basis of a semi-quantitative test against casein or gelative by cell-free filtrates, they found strains of *C. lipolytica, Aureobasidium pullulans* and *Cephalosporium* sp. invariably positive. According to Maddox and Hough,[193] at least four proteinases are secreted by growing cells of *Sacch. uvarum*. The secretion of active proteinases accompanied growth on casein but not casein hydrolysate. After partial purification two of the enzyme fractions still contained substantial amounts of carbohydrate (22% and 38%) and may be glycoproteins.[193] The relative amounts of secreted and retained proteinases, and their relationships (if any) to the intracellular proteinases, have not been elucidated.

Provided that a given substrate can penetrate the cell wall one would suppose, from the evolutionary point of view, that retention of the specific hydrolytic enzyme in the periplasmic space constitutes an advantage. For the breakdown of larger polymeric nutrients the cell has no alternative but to secrete the requisite enzymes into the medium. A comparison among effective sizes (hydrodynamic radii) of secreted and retained enzymes, together with accessible space (see Volume I, Chapter 3) measurements on the parent cells, would be potentially informative.

VIII. BIOSYNTHESIS OF CELL ENVELOPE ENZYMES

A. Repression, Derepression, and Induction

Certain yeast enzymes are constitutive in the sense that they are produced irrespective of the chemical environment. On the other hand, the degrees of production of other enzymes are influenced by the concentration of potential substrates or enzymatic products. In the negative sense this effect is called repression, i.e., a decreased synthesis of the enzyme in question in the presence of a certain concentration of metabolite. A specific example is the low titer of acid phosphatase achieved by yeast cells grown in the presence of normal amounts of orthophosphate. However, below a threshold concentration of orthophosphate the enzyme is synthesized in substantial amounts, i.e., the synthetic system has been derepressed.

In the positive sense we have the phenomenon of induction. An inducible enzyme is one that is only synthesized in response to the presence of a particular compound or compounds, i.e., so-called inducers. Included in the category of inducible systems are certain permeases associated with the plasma membrane (see Volume I, Chapter 4).

The mechanisms of repression, derepression, and induction are currently of great interest. However, a discussion of the molecular biology of these phenomena is beyond the scope of this volume. From the point of view of the enzymologist, the influences of certain growth media constituents will be of immediate practical import. Thus the selection of a growth medium may greatly influence the amount of enzyme per cell and hence the enthusiasm for undertaking certain biochemical studies. To this end some examples and literature references are assembled in Table 5.

The concentrations (Table 5) effecting repression are only approximate because of fluxes due to metabolism. More accurate indications on growing cells could be obtained with continuous cultures in which the effector concentration is held constant. Thus Davies[199] has shown that β-fructofuranosidase synthesis in *K. fragilis* is repressed by

Table 5
SELECTED CELL ENVELOPE ENZYMES KNOWN TO BE SUBJECT TO REPRESSION OR INDUCTION

Enzyme	Species	Compound	Concentration	Ref.
		I. Repression		
β-Fructofuranosidase	*Saccharomyces cerevisiae*	Glucose	28—56 m*M*	194—196
	Kluyveromyces fragilis	Glucose or fructose	50 m*M*	197
Acid phosphatase	*Saccharomyces cerevisiae*	Orthophosphate	1 m*M*	96
	Saccharomyces bisporus	Orthophosphate	~4 m*M*	198
	Saccharomyces rouxii	Orthophosphate	2 m*M*	7
	Schizosaccharomyces pombe	Orthophosphate	~4 m*M*	79
Asparaginase II	*Saccharomyces cerevisiae*	Nitrogen catabolites	2 m*M*	177
		II. Induction		
α-Galactosidase	*Saccharomyces uvarum*	Galactose, melibiose, or raffinose		113 114
N-Acetylglucosamine- Kinase Deacetylase Deaminase	*Candida albicans*	*N*-acetylglu-cosamine		182
Glucamylase	*Saccharomyces diastaticus*	Starch		184
Plasma membrane transport systems			See Volume I, Chapter 4	

much lower concentrations than indicated in Table 5 provided the concentration is sustained (i.e., chemostat conditions). Mutants of *Sacch. cerevisiae* have been selected on the basis of tolerating higher concentrations of zymohexose (glucose, fructose, and mannose) during β-fructofuranosidase synthesis.[47] There is a potential danger in using mutants in that one example was found to possess a bizarre cell envelope.[72] It is also worth noting in passing that cytoplasmic β-fructofuranosidase is not subject to glucose repression.[47]

Orthophosphate is invariably a repressor of acid phosphatase synthesis in yeasts. Apart from the species listed in Table 5, one can mention *C. utilis,*[86] *Rh. rubra,*[80] *C. albicans,*[108] and *H. holstii*[104] which require a depleted phosphate medium (concentrations not reported) for acid phosphatase synthesis. The major source of orthophosphate in a complex growth medium is usually yeast extract. Concentrated stock solutions of this constituent can be pretreated with magnesia to deplete phosphate. Because phosphate is essential for growth, it is worthwhile to define the preculture conditions and the carry-over of phosphate in the experimental inoculum.[7]

To avoid catabolite repression of β-fructofuranosidase synthesis one can use a slow but continuous addition of glucose or sucrose to the growing culture.[31] This technique has been practiced since at least 1925 (Reference 9). Modern culture apparatus provides for facile additions under sterile conditions by using peristaltic pumps.

Aminopeptidase II, which is claimed to be extracytoplasmic, is apparently not repressed by glucose and is not influenced by the nitrogen content of the medium.[120] This is somewhat surprising; the more so because aminopeptidase I (internal) is repressed by high concentrations of glucose and is influenced by the nitrogen nutrition of the cell.

B. Translocation

It is generally accepted that all cell envelope enzymes are synthesized within the protoplasm because of the demonstrated synthetic ability of isolated protoplasts. The mechanisms by which these extracytoplasmic enzymes are transported to their final locales come under the heading of translocation. The plasma membrane must be crossed and it is not unreasonable to suspect that structural changes in the enzyme may occur during transit. These changes may include covalent modifications (e.g., addition of sugar residues to glycoproteins or processing of zymogen forms) as well as conformational transitions culminating in the final structures of the macromolecules. Although bacterial and higher eukaryotic examples of such modifications are on hand, these events have yet to be described in yeasts.

The periplasmic space enzymes have rightly been the most attractive for study. On the other hand, the mechanisms for the supposed insertion of β-glucanases into existing cell walls can only be contemplated at present.

Three working hypotheses have received some attention and these three are diagrammatically summarized in Figure 4. The first such model (Figure 4A) supposes that prepackaged enzymes are delivered via vesicles whose limiting membrane fuses with the plasma membrane. This is the model[200] for the secretion of zymogen granules of the mammalian pancreas. Another well-documented system is that of mucocyst discharge by the ciliate protozoan *Tetrahymena pyriformis*.[201,202] Moor[203] employed freeze-fracture techniques in a microscopic study of membrane systems during yeast cell budding. He demonstrated production of vesicles which were thought to initiate with the endoplasmic reticulum and clearly became concentrated in the bud initial. It was originally speculated[203] that these vesicles might contain protein-disulfide reductase, a key character in the plasticity-increase hypothesis originally championed by Nickerson and colleagues.[204] Matile et al.[131] subsequently suggested that similar vesicles contain β-glucanases that might be involved in local cell wall modification. Several investigators have reported on vesicles associated with bud initials, although agreement on the distribution of enzymes within them has been less than satisfying.

Novick and Schekman[205] generated a temperature-sensitive mutant of *Sacch. cerevisiae* whose secretion of acid phosphatase and β-fructofuranosidase into the periplasmic space is impaired at 37°C but not at 24°C. At the nonpermissive temperature, protoplasts of the mutant showed a great proliferation of vesicles that were acid phosphatase positive by cytochemistry. A related mutant, built with a defective structural gene for acid phosphatase, showed proliferation of vesicles that were in this case negative by lead capture cytochemistry. The nature of the presumptive lesion is not known, but supposedly does not allow productive membrane fusion between vesicles and plasma membrane.

The second model (Figure 4B) suggests that the biosynthetic apparatus (ribosomes and messenger RNA) is intimately associated with the plasma membrane. The actual association may depend upon the appearance of a signal sequence at the front of the growing polypeptide product.[206] Evidence to support extrusion of some polypeptides through a plasma membrane comes from studies with bacteria. Thus *Bacillus amyloliquefaciens* protoplasts secrete α-amylase as well as a protease which, during the secretion process, are temporarily more susceptible to proteolytic inactivation by external trypsin or chymotrypsin.[207] The organism's own protease, if completely folded, will also attack the above-mentioned nascent polypeptides. In this particular bacterial

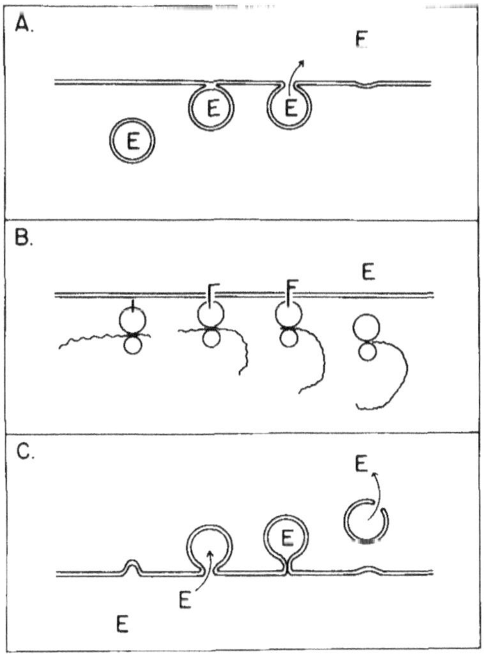

FIGURE 4. Schematic representation of three working hypotheses for translocation of enzymes into the periplasmic space. In each case the temporal sequence is from left to right. The enzyme (E) crosses the plasma membrane (double track) from the cytoplasmic side (down) to the periplasmic space side (up). The three schemes are discussed in the text.

system no significant amounts of α-amylase or protease are ever found inside secreting cells, and no immunologically cross-reactive material is detected in the cytoplasm.

Lazdunski et al.[208] have shown that a local anesthetic (procaine) can interact with *Escherichia coli* membranes and so inhibit the export of periplasmic proteins. Immature (longer) precursors accumulate, supposedly because the signal peptidase[209] is inhibited by procaine.

Holbein and Kidby[210] are of the opinion that extrusion of a nascent β-fructofuranosidase polypeptide occurs with *Sacch. cerevisiae* protoplasts. Their data purport to demonstrate increased susceptibility to trypsin digestion during the extrusion process. However, the plasma membrane is damaged by their treatments, as judged by increased susceptibility to lysis, and other interpretations are possible.

A third working hypothesis (Figure 4C) comes mainly from our experience with the β-fructofuranosidase of *Sacch. rouxii* and the unusual periplasmic ultrastructure displayed by this species. Growing cells of *Sacch. rouxii* do not ferment sucrose, but, after considerable aging of the culture, the disaccharide is assimilated as well as is glucose or fructose.[41] The change is due to the expression of a physically cryptic β-fructofuranosidase.[6] Young cells exhibit this enzyme activity if pretreated by any one of a variety of membrane-disrupting perturbations. Those treatments that also break the cell wall concomitantly release the enzyme to the medium, whereas aged cells, and cells treated with lipid solvents, retain the expressed enzyme. The enzyme of young cells is still cryptic in naked protoplasts and, upon lysis by dilution, they subsequently release and express the activity.[6,26] Thus there is good evidence to support the contention that the cryptic enzyme is surrounded by the plasma membrane or some other membrane that prohibits access to substrates (sucrose or raffinose).

The ultrastructure of *Sacch. rouxii* is of interest in that young cells display a propensity for periplasmic bodies (see Volume II, Chapter 8, Figure 7), membrane-surrounded vesicles which originate from the plasma membrane[26,211] and are much less prevalent in sections of mature cells.[26] Periplasmic bodies have been observed in three different strains of *Sacch. rouxii*[211] and after three different fixation regimens.[26] On the other hand, periplasmic bodies are much less prevalent in other *Saccharomyces* species. The admittedly circumstantial evidence that periplasmic bodies are present in *Sacch. rouxii* cells when their β-fructofuranosidase is cryptic prompted us to suggest that the unusual organelle might be involved in the translocation of periplasmic enzymes. Evidence to support or dismiss this hypothesis would follow from isolation, or at least enrichment, of periplasmic bodies and a critical comparison of specific activities. To date this has not been accomplished, and it is supposed that a contributing if not prohibitive problem in this methodology is that most of the periplasmic bodies remain attached to the plasma membrane through a thin pedicel.[26] This would explain why a variety of gradient centrifugation procedures has failed to separate periplasmic bodies from protoplasts. Attempts to dislodge these vesicles with detergents or mild sonication have resulted in lysis of protoplasts as well.

An organelle of comparative interest is the lomasome which has been described in filamentous fungi.[212] Ghosh et al.[213] have discussed a possible relationship between penicillinase secretory structures and mesosomes in *Bacillus licheniformis* and in other organisms. The propensity for periplasmic bodies in *Sacch. rouxii* recommends protoplasts of that species for further study. But whether these structures have anything to do with cryptic β-fructofuranosidase or its translocation is still unknown. One piece of evidence that already negates the generality of this hypothesis is that the acid phosphatase of this species is always expressed, i.e., like that in other *Saccharomyces* species.

IX. CONCLUDING REMARKS

Considering all of the enzymes that are either known, or reasonably anticipated, to be associated with yeast cell envelopes it is clear that the majority of extant studies has been devoted to hydrolytic enzymes. This is not surprising because of the relatively large concentrations of β-fructofuranosidase and acid phosphatase in particular, their stability, and their ease of assay. An emerging concept is that periplasmic space enzymes, once generated, are probably not under sophisticated control mechanisms. The syntheses of periplasmic space enzymes are subject to repression, but, because they are simply restrained by the cell wall, it is not anticipated that they will be greatly influenced by surrounding compounds.

On the other hand, the enzymes of the plasma membrane are only beginning to be studied in any detail, and already rather complicated interactions with positive and negative effectors are being proposed. It must also be anticipated that the lipid environment may exert significant influence on various membrane enzymes. A precedent here is the well-documented influence of the hydrophobic environment on glucose-6-phosphatase within mammalian endoplasmic reticulum.[214]

The synthetic enzymes that are imagined to play on existing polymers in the growing cell wall demand further study, but obviously they are not easy experimental subjects. The procedural problems therein may be compounded by the indirect effects of disruption on the matrix of the wall itself.

The capsular materials of *Cryptococcus* species have been analyzed, but virtually nothing is known about the enzymes involved. The enzymology associated with the cell envelopes of dimorphic yeast-like organisms must surely contribute to yeast phase-mycelial phase interconversions which are important to the etiology of pathogens such as *Candida albicans* and *Histoplasma capsulatum*. In light of the increased incidence of fungal diseases in man we may anticipate vigorous activity in these relatively unexplored areas.

X. ACKNOWLEDGMENT

Literature retrieval in connection with this chapter was greatly facilitated by the assistance of Brenda Johnson.

REFERENCES

1. **Conway, E. J. and Downey, M.**, An outer metabolic region of the yeast cell, *Biochem. J.*, 47, 347, 1950.
2. **Willstätter, R. and Racke, F.**, Zur Kenntnis des Invertins, *Justus Liebigs Ann. Chem.*, 427, 111, 1921.
3. **Schmalreck, A. F. and Teuber, M.**, Effect of chemical modification by (di) imidoesters on cells and cell envelope components of *Escherichia coli* and *Salmonella typhimurium*, *Microbios*, 17, 93, 1976.
4. **Dixon, M. and Webb, E. C.**, *Enzymes*, Academic Press, New York, 1958, 95.
5. **Demis, D. J., Rothstein, A., and Meier, R.**, The relationship of the cell surface to metabolism. X. The location and function of invertase in the yeast cell, *Arch. Biochem. Biophys.*, 48, 55, 1954.
6. **Arnold, W. N.**, Expression of cryptic β-fructofuranosidase in *Saccharomyces rouxii*, *J. Bacteriol.*, 120, 886, 1974.
7. **Arnold, W. N. and Garrison, R. G.**, An Fe^{3+}-activated acid phosphatase in *Saccharomyces rouxii*, *J. Biol. Chem.*, 254, 4919, 1979.
8. **Holzer, H.**, Regulatory role of proteinases and proteinase inhibitors in yeast, in *Mechanisms of Action and Regulation of Enzymes*, Proc. 9th F.E.B.S. Meeting, Keleti, T., Ed., Akademikai Kiado, Budapest, 1975, 181.
9. **Neuberg, C. and Roberts, I. S.**, Invertase, *Sugar Res. Found. Sci. Rep. Ser.*, 4, 1946.
10. **Neuberg, C. and Mandl, I.**, Invertase, in *The Enzymes*, Vol. 1 (Part 1), Sumner, J. B. and Myrbäck, K., Eds., Academic Press, New York, 1950, chap. 14.
11. **Gottschalk, A.**, Principles underlying enzyme specificity in the domain of carbohydrates, *Adv. Carbohydr. Chem.*, 5, 49, 1950.
12. **Hestrin, S., Feingold, D. S., and Schramm, M.**, Hexoside hydrolases, in *Methods in Enzymology*, Vol. 1, Colowick, S. P. and Kaplan, N. O., Eds., Academic Press, New York, 1955, chap. 29.
13. **Bacon, J. S. D.**, Methods for measuring transglycosylase activity of invertases, in *Methods in Enzymology*, Vol. 1, Colowick, S. P. and Kaplan, N. O., Eds., Academic Press, New York, 1955, chap. 30.
14. **Myrbäck, K.**, Invertases, in *The Enzymes*, Vol. 4, 2nd ed., Boyer, P. D., Lardy, H., and Myrbäck, K., Eds., Academic Press, New York, 1960, chap. 22.
15. **Nisizawa, K. and Hashimoto, Y.**, Glycoside hydrolases and glycosyl transferases, in *The Carbohydrates, Chemistry and Biochemistry*, Vol. 2A, 2nd ed., Pigman, W. and Horton, D., Eds., Academic Press, New York, 1970, chap. 33.
16. **Lampen, J. O.**, Yeast and neurospora invertases, in *The Enzymes*, Vol. 5, 3rd ed., Boyer, P. D., Lardy, H., and Myrbäck, K., Academic Press, New York, 1971, chap. 10.
17. **Goldstein, A. and Lampen, J. O.**, β-D-Fructofuranoside fructohydrolase from yeast, in *Methods in Enzymology*, Vol. 42, Wood, W. A., Ed., Academic Press, New York, 1975, chap. 76.
18. **Koshland, D. E., Jr. and Stein, S. S.**, Correlation of bond breaking with enzyme specificity cleavage point of invertase, *J. Biol. Chem.*, 208, 139, 1954.
19. **Bacon, J. S. D. and Edelman, J.**, The action of invertase preparations, *Arch. Biochem. Biophys.*, 28, 467, 1950.

20. **Blanchard, P. H. and Albon, N.,** The inversion of sucrose; a complication, *Arch. Biochem. Biophys.*, 29, 220, 1950.
21. **Bacon, J. S. D.,** The oligosaccharides produced by the action of yeast invertase preparations on sucrose, *Biochem, J.*, 57, 320, 1954.
22. **Andersen, B.,** The transferring activity of β-fructofuranosidase: formation of two disaccharides from fructose, *Acta. Chem. Scand.*, 21, 828, 1967.
23. **Sumner, J. B.,** The estimation of sugar in diabetic urine, using dinitrosalicylic acid, *J. Biol. Chem.*, 62, 287, 1924 to 1925.
24. **Arnold, W. N.,** β-Fructofuranosidase from grape berries, *Biochim. Biophys. Acta*, 110, 134, 1965.
25. **Arnold, W. N.,** *p*-Toluenethiol as an initiator of autolysis in bakers' yeast, *J. Bacteriol.*, 109, 949, 1972.
26. **Arnold, W. N. and Garrison, R. G.,** Isolation and characterization of protoplasts from *Saccharomyces rouxii*,, *J. Bacteriol.*, 137, 1386, 1979.
27. **Bestic, P. B. and Arnold, W. N.,** Linear transformation of standard curves for yeast turbidity, *Appl. Environ. Microbiol.*, 32, 640, 1976.
28. **Arnold, W. N.,** Heat inactivation kinetics for yeast β-fructofuranosidase. A polydisperse system, *Biochim. Biophys. Acta*, 178, 347, 1969.
29. **Phaff, H. J.,** Structure and biosynthesis of the yeast cell envelope, in *The yeasts*, Vol. 2, Rose, A. H. and Harrison, J. S., Eds., Academic Press, London, 1971, chap. 5.
30. **Burger, M., Bacon, E. E., and Bacon, J. S. D.,** Some observations on the forms and location of invertase in the yeast cell, *Biochem, J.*, 78, 504, 1961.
31. **Arnold, W. N.,** Location of acid phosphatase and β-fructofuranosidase within yeast cell envelopes, *J. Bacteriol.*, 112, 1346, 1972.
32. **Arnold, W. N.,** The structure of the yeast cell wall. Solubilization of a marker enzyme, β-fructofuranosidase, by the autolytic enzyme system, *J. Biol. Chem.*, 247, 1161, 1972.
33. **Lampen, J. O.,** External enzymes of yeast: their nature and formation, *Antonie van Leeuwenhoek, J. Microbiol. Serol.*, 34, 1, 1968.
34. **Nagasaki, S., Neumann, N. P., Arnow, P., Schnable, L. D., and Lampen, J. O.,** An enzyme which degrades the walls of living yeast, *Biochem. Biophys. Res. Commun.*, 25, 158, 1966.
35. **Sutton, D. D. and Lampen, J. O.,** Localization of sucrose and maltose fermenting systems in *Saccharomyces cerevisiae, Biochim. Biophys. Acta*, 56, 303, 1962.
36. **Eddy, A. A. and Williamson, D. H.,** Formation of aberrant cell walls and of spores by the growing yeast protoplast, *Nature (London)*, 183, 1101, 1959.
37. **McMurrough, I. and Rose, A. H.,** Effect of growth rate and substrate limitation on the composition and structure of the cell wall of *Saccharomyces cerevisiae, Biochem, J.*, 105, 189, 1967.
38. **Preiss, J. W.,** The localization of invertase in the yeast cell with low voltage electrons, *Arch. Biochem. Biophys.*, 75, 186, 1958.
39. **Wickerham, L. J.,** Evidence of the production of extracellular invertase by certain strains of yeasts, *Arch. Biochem. Biophys.*, 76, 439, 1958.
40. **Dworschack, R. G. and Wickerham, L. J.,** Production of extracellular invertase by the yeast *Saccharomyces uvarum* NRRL Y-972, *Arch. Biochem. Biophys.*, 76, 449, 1958.
41. **Pappagianis, D. and Phaff, H. J.,** Delayed fermentation of sucrose by certain haploid species of *Saccharomyces, Antonie van Leeuwenhoek, Microbiol. Serol., J.*, 22, 353, 1956.
42. **Kidby, D. K. and Davies, R.,** Invertase and disulphide bridges in the yeast cell wall, *J. Gen. Microbiol.*, 61, 327, 1970.
43. **Kidby, D. K. and Davies, R.,** Thiol induced release of invertase from cell walls of *Saccharomyces fragilis, Biochim. Biophys. Acta*, 201, 261, 1970.
44. **Schwencke, J., Farias, G., and Rojas, M.,** The release of extracellular enzymes from yeast by osmotic shock, *Eur. J. Biochem.*, 21, 137, 1971.
45. **Friis, J. and Ottolenghi, P.,** Localization of invertase in a strain of yeast, *C. R. Trav. Lab. Carlsberg*, 31, 259, 1959.
46. **Gascon, S. and Lampen, J. O.,** Purification of the internal invertase of yeast, *J. Biol. Chem.*, 243, 1567, 1968.
47. **Gallilli, G. and Lampen, J. O.,** Large and small invertases and the yeast cell cycle pattern of synthesis and sensitivity to tunicamycin, *Biochim. Biophys. Acta*, 475, 113, 1977.
48. **Crane, R. K.,** Digestion and absorption of carbohydrates, in *Carbohydrate Metabolism and Its Disorders*, Vol. 1, Dickens, F., Randle, P. J., and Whelan, W. J., Eds., Academic Press, London, 1968, chap. 2.
49. **Willstätter, R.,** *Untersuchungen über Enzyme*, Vol. 1, Springer-Verlag, Berlin, 1928, 535 to 856.
50. **Sumner, J. B. and O'Kane, D. J.,** The chemical nature of yeast saccharase, *Enzymologia*, 12, 251, 1948.
51. **Fischer, E. H. and Kohtes, L.,** Purification de l'invertase de levure, *Helv. Chim. Acta.*, 34, 1123, 1951.

52. **Andersen, B.**, Isolation of β-fructofuranosidase from yeast by ion exchange chromatography on diethylaminoethyl cellulose, *Acta. Chem. Scand.*, 14, 1849, 1960.
53. **Myrbäck, K. and Schilling, W.**, Studies on yeast β-fructofuranosidase. XVI. Partial purification, *Enzymologia,* 29, 306, 1965.
54. **Neumann, N. P. and Lampen, J. O.**, Purification and properties of yeast invertase, *Biochemistry*, 6, 468, 1967.
55. **Andersen, B. and Jorgensen, O. S.**, Purification of external invertase from brewers' yeast, *Acta. Chem. Scand.*, 23, 2270, 1969.
56. **Trimble, R. B. and Maley, F.**, Subunit structure of external invertase from *Saccharomyces cerevisiae*, *J. Biol. Chem.*, 252, 4409, 1977.
57. **Sizer, I. W.**, Sucrose inversion by invertase as a function of temperature, *Enzymologia,* 2, 215, 1938.
58. **Greiling, H., Vógele, P., Kisters, R., and Ohlenbusch, H. D.**, Über eine kohlenhydrat Protein-Binding in der β-Fructofuranosidase, *Hoppe-Seyler's Z. Physiol. Chem.*, 350, 517, 1969.
59. **Arnold, W. N.**, Chromatography of a polydisperse enzyme, yeast β-fructofuranosidase, on DEAE-cellulose, *Anal. Lett.*, 2, 547, 1969.
60. **Berggren, B.**, Studies on yeast β-fructofuranosidase (invertase). XXIII. Studies on the heterogeneity of the enzyme and its liberation from the yeast, *Ark., Kemi,* 29, 117, 1968.
61. **Colonna, W. J., Cano, F. R., and Lampen, J. O.**, Microheterogeneity in yeast invertase, *Biochim. Biophys. Acta*, 386, 293, 1975.
62. **Neumann, N. P. and Lampen, J. O.**, The glycoprotein structure of yeast invertase, *Biochemistry,* 8, 3552, 1969.
63. **Sentandreu, R. and Northcote, D. H.**, The structure of a glycopeptide isolated from the yeast cell wall, *Biochem, J.*, 109, 419, 1968.
64. **Gascon, S., Neumann, N. P., and Lampen, J. O.**, Comparative study of the properties of the purified internal and external invertases from yeast, *J. Biol. Chem.*, 243, 1573, 1968.
65. **von Euler, H. and Laurin, I.**, Über die Temperaturempfindlichkeit der Saccharase (Invertase), *Z. Physiol. Chem.*, 108, 64, 1919.
66. **Chu, F. K., Trimble, R. B., and Maley, F.**, The effect of carbohydrate depletion on the properties of yeast external invertase, *J. Biol. Chem.*, 253, 8691, 1978.
67. **Smith, W. L. and Ballou, C. E.**, Immunochemical characterization of the mannan component of the external invertase of *Saccharomyces cerevisiae*, *Biochemistry*, 13, 355, 1974.
68. **Nelson, J. M. and Kerr, R. W.**, Uniformity in invertase action. III. The stability of the enzyme, *J. Biol. Chem.*, 59, 495, 1924.
69. **Ottolenghi, P.**, A comparison of five genetically distinct invertases from *Saccharomyces*, *Eur. J. Biochem.*, 18, 544, 1971.
70. **Braun, H.**, External yeast β-fructofuranosidase affinity labelling of the active site, *Biochim. Biophys Acta*, 452, 452, 1976.
71. **Strumeyer, D. H. and Malin, M. J.**, Resistance of extracellular yeast invertase and other glycoproteins to denaturation by tannins, *Biochem, J.*, 118, 899, 1970.
72. **Ghosh, B. K., Montenecourt, B., and Lampen, J. O.**, Abnormal cell envelope ultrastructure of a *Saccharomyces* mutant with invertase formation resistant to hexoses, *J. Bacteriol.*, 116, 1412, 1973.
73. **Heredia, C. F., Yen, F., and Sols, A.**, Role and formation of the acid phosphatase in yeast, *Biochem. Biophys. Res. Commun.*, 10, 14, 1963.
74. **Gomori, G.**, Distribution of acid phosphatase in tissue sections, *Arch. Pathol.*, 32, 189, 1941.
75. **Boer, P. and Steyn-Parve, E. P.**, Further studies on the mechanism of action of an acid phosphatase from bakers' yeast *(Saccharomyces cerevisiae)*, *Biochim. Biophys. Acta*, 206, 281, 1970.
76. **Boer, P. and Steyn-Parve, E. P.**, Some kinetic aspects of the mechanism of action of an acid phosphatase from bakers' yeast *(Saccharomyces cerevisiae)*, *Biochim. Biophys. Acta*, 171, 360, 1969.
77. **Albers, H., Büsing, G., and Schudt, G.**, Über die aktive Gruppe der sauren Hefephosphatase, *Enzymologia,* 30, 149, 1966.
78. **Dibenedetto, G. and Mura, U.**, Some kinetic aspects of the mechanism of hydrolysis of phosphoric acid esters by non-specific acid phosphatase from *Schizosaccharomyces pombe*, *Biochim. Biophys. Acta*, 522, 122, 1978.
79. **Dibenedetto, G.**, Acid phosphatase in *Schizosaccharomyces pombe*. I. Regulation and preliminary characterization, *Biochim. Biophys. Acta*, 286, 363, 1972.
80. **Watorek, W., Morawiecka, B., and Korczak, B.**, Acid phosphatase of the yeast *Rhodotorula rubra.* Purification and properties of the enzyme, *Arch. Biochim. Pol.*, 24, 153, 1977.
81. **Tonino, G. J. M. and Steyn-Parve, E. P.**, Localization of some phosphatases in yeast, *Biochim. Biophys. Acta*, 67, 453, 1963.
82. **Rothstein, A. and Meier, R.**, The relationship of the cell surface to metabolism. I. Phosphatases in the cell surface of living yeast cells, *J. Cell. Comp. Physiol.*, 32, 77, 1948.

83. **Albers, H. and Albers, E.,** Über die Phosphatasen der Hefe. 4. Mittellung zur Kenntnis der Phosphatasen, *Hoppe-Seyler's Z. Physiol. Chem.*, 235, 47, 1935.
84. **Nurminen, T., Oura, E., and Suomalainen, H.,** The enzymic composition of the isolated cell wall and plasma membrane of bakers' yeast, *Biochem, J.*, 116, 61, 1970.
85. **Weimberg, R. and Orton, W. L.,** Elution of acid phosphatase from the cell surface of *Saccharomyces mellis* by potassium chloride, *J. Bacteriol.*, 90, 82, 1965.
86. **Rautanen, N. and Kaerkkaeinen, V.,** On the phosphatase activity of low-phosphorus *Torulopsis utilis, Acta. Chem. Scand.*, 5, 1216, 1951.
87. **Schmidt, G., Bartsch, G., Laumont, M. C., Herman, T., and Liss, M.,** Acid phosphatase of bakers' yeast: an enzyme of the external cell surface, *Biochemistry*, 2, 126, 1963.
88. **Suomalainen, H., Linko, M., and Oura, E.,** Changes in the phosphatase activity of bakers' yeast during the growth phase and location of the phosphatases in the yeast cell, *Biochim. Biophys. Acta*, 37, 482, 1960.
89. **Boer, P. and Steyn-Parve, E. P.,** Isolation and purification of an acid phosphatase from bakers' yeast (*Saccharomyces cerevisiae*), *Biochim. Biophys. Acta*, 128, 400, 1966.
90. **Chen, P. S., Toribara, T. Y., and Warner, H.,** Microdetermination of phosphorous, *Anal. Chem.*, 11, 1756, 1956.
91. **Taussky, H. H. and Shorr, E.,** A micro colorimetric method for the determination of inorganic phosphorous, *J. Biol. Chem.*, 202, 675, 1953.
92. **Arnold, W. N. and Lacy, J. S.,** Permeability of the cell envelope and osmotic behavior in *Saccharomyces cerevisiae, J. Bacteriol.*, 131, 564, 1977.
93. **Suomalainen, H., Nurminen, T., and Oura, E.,** Leakage of some enzymes and cofactors from the cell during preparation of protoplasts from bakers' yeast, *Arch. Biochem. Biophys.*, 118, 219, 1967.
94. **McLellan, W. L. and Lampen, J. O.,** The acid phosphatase of yeast. Localization and secretion by protoplasts, *Biochim. Biophys. Acta*, 67, 324, 1963.
95. **Weimberg, R. and Orton, W. L.,** Evidence for an exocellular site for the acid phosphatase of *Saccharomyces mellis, J. Bacteriol.*, 88, 1743, 1964.
96. **van Rijn, H. J. M., Boer, P., and Steyn-Parve, E. P.,** Biosynthesis of acid phosphatase of bakers' yeast. Factors influencing its production by protoplasts and characterization of the secreted enzyme, *Biochim. Biophys. Acta*, 268, 431, 1972.
97. **Boer, P., van Rijn, H. J. M., Reinking, A., and Steyn-Parve, E. P.,** Biosynthesis of acid phosphatase of bakers' yeast. Characterization of a protoplast-bound fraction containing precursors of the exoenzyme, *Biochim. Biophys. Acta*, 377, 331, 1975.
98. **Gunther, T., Kattner, W., and Merker, H. J.,** Über das Verhalten und die Lokalisation der Sauren Phosphatase von Hefezellen bei Repression und Derepression, *Exp. Cell Res.*, 45, 133, 1966.
99. **Bauer, H. and Sigarlakie, E.,** Cytochemistry on ultrathin frozen sections of yeast cells. Localization of acid and alkaline phosphatase, *J. Microsc.*, 99 (2), 205, 1973.
100. **Linnemans, W. A. M., Boer, P., and Elbers, P. F.,** Localization of acid phosphatase in *Saccharomyces cerevisiae*: a clue to cell wall formation, *J. Bacteriol.*, 131, 638, 1977.
101. **Pfeifer, U., Poehlmann, E., and Witschel, H.,** Kinetics of the accumulation of lead phosphate in acid phosphatase staining, in *Electron Microscopy and Cytochemistry,* Weisse, E., Daems, W. T., Molenaar, I., and van Duijn, P., Eds., North-Holland, Amsterdam, 1973.
102. **Clarkson, T. W. and Kench, J. E.,** Uptake of lead by human erythrocytes *in vitro, Biochem, J.*, 69, 432, 1958.
103. **Nurminen, T., Oura, E., and Suomalainen, H.,** The enzymic composition of the isolated cell wall and plasma membrane of bakers' yeast, *Biochem, J.*, 116, 61, 1970.
104. **San Blas, G. and Cunningham, W. L.,** Production of an exocellular acid phosphatase by the yeast *Hansenula holstii* NCYC 560, *Biochim. Biophys. Acta*, 343, 208, 1974.
105. **Mahvi, T., Spicer, S. S., and Wright, N. J.,** Cytochemistry of acid mucosubstance and acid phosphatase in *Cryptococcus neoformans, Can. J. Microbiol.*, 20, 833, 1974.
106. **de Nollin, S., Thorne, F., and Borgers, M.,** Enzyme cytochemistry of *Candida albicans, J. Histochem. Cytochem.*, 23, 758, 1975.
107. **Dibenedetto, G. and Cozzani, I.,** Nonspecific acid phosphatase from *Schizosaccharomyces pombe.* Purification and physical chemical properties, *Biochemistry,* 14, 2847, 1975.
108. **Odds, F. C. and Hierholzer, J. C.,** Purification and properties of a glycoprotein acid phosphatase from *Candida albicans, J. Bacteriol.*, 114, 257, 1973.
109. **Schurr, A. and Yagil, E.,** Regulation and characterization of acid and alkaline phosphatases in yeast, *J. Gen. Microbiol.*, 65, 291, 1971.
110. **Shetlar, M. R., Foster, J. V., and Everett, M. R.,** Determination of serum polysaccharides by the tryptophane reaction, *Proc. Soc. Exp. Biol. Med.*, 67, 125, 1948.
111. **Mildner, P., Ries, B., and Barbaric, S.,** Acid phosphatase and adenosine triphosphatase activities in the cell wall of bakers' yeast, *Biochim. Biophys. Acta*, 391, 67, 1975.

112. **Chen, T. T., Boyer, F. W., Cetorelli, J. T., Pollard, W. E., and Roberts, R. M.,** Purification and properties of a progesterone-induced basic glycoprotein from the uterine fluids of pigs, *J. Biol. Chem.*, 248, 8560, 1973.
113. **Sols, A. and de la Fuente, G.,** Transport and hydrolysis in the utilization of oligosaccharides by yeasts, in *Membrane transport and metabolism, Proc. Symp. Prague,* Academic Press, London, 1960, 361.
114. **Friis, J. and Ottolenghi, P.,** Localization of melibiase in a strain of yeast, *C. R. Trav. Lab. Carlsberg,* 31, 272, 1960.
115. **Arnold, W. N.,** Trehalose assimilation and turnover in *Torulopsis glabrata, Current Microbiol.*, 2, 109, 1979.
116. **Arnold, W. N. and McLellan, M. N.,** Trehalose and glycogen levels during the initial stages of growth of *Candida albicans, Physiol. Chem. Phys.*, 7, 369, 1975.
117. **Koytk, A. and Michaljanicova, D.,** Uptake of trehalose by *Saccharomyces cerevisiae, J. Gen. Microbiol.*, 110, 323, 1979.
118. **Kaplan, J. G. and Tacreiter, W.,** The β-glucosidase of the yeast cell surface, *J. Gen. Physiol.*, 50, 9, 1966.
119. **Kaplan, J. G.,** An inducible system for the hydrolysis and transport of β-glucosides in yeast. I. Characteristics of the β-glucosidase activity of intact and lysed cells, *J. Gen. Physiol.*, 48, 873, 1965.
120. **Frey, J. and Röhm, K. H.,** Subcellular localization and levels of aminopeptidases and dipeptidases in *Saccharomyces cerevisiae, Biochim. Biophys. Acta*, 527, 31, 1978.
121. **Marder, R., Becker, J. M., and Naider, F.,** Peptide transport in yeast: utilization of leucine- and lysine-containing peptides by *Saccharomyces cerevisiae, J. Bacteriol.*, 131, 906, 1977.
122. **Lewis, P. R. and Knight, D. P.,** *Staining methods for sectioned material,* North-Holland, Amsterdam, 1977, 112.
123. **Pease, D. C.,** Glycol methacrylate copolymerized with glutaraldehyde and urea as an embedment retaining lipids, *J. Ultrastruct. Res.*, 45, 124, 1973.
124. **Barras, D. R.,** A β-glucan endohydrolase from *Schizosaccharomyces pombe* and its role in cell wall growth, *Antonie van Leeuwenhoek; Microbiol. Serol. J.*, 38, 65, 1972.
125. **Megnet, R.,** Effect of 2-deoxyglucose on *Schizosaccharomyces pombe, J. Bacteriol.*, 90, 1032, 1965.
126. **Johnson, B. F.,** Lysis of yeast cell walls induced by 2-deoxyglucose at their sites of glucan synthesis, *J. Bacteriol.*, 95, 1169, 1968.
127. **Biely, P., Farkaš, V., and Krátký, Z.,** Some aspects of wall formation in *Schizosaccharomyces pombe.* A study with 2-deoxyglucose, *Biologia*, 29, 919, 1974.
128. **Krátký, Z., Biely, P., and Bauer, S.,** Mechanism of 2-deoxy-D-glucose inhibition of cell wall polysaccharide and glycoprotein biosynthesis in *Saccharomyces cerevisiae, Eur. J. Biochem.*, 54, 459, 1975.
129. **Fleet, G. H. and Phaff, H. J.,** Glucanases in *Schizosaccharomyces.* Isolation and properties of the cell wall associated (1-3)-glucanases, *J. Biol. Chem.*, 249, 1717, 1974.
130. **Bacon, J. S. D., Jones, D., Farmer, V. C., and Webley, D. M.,** The occurrence of α(1 → 3)-glucan in *Cryptococcus, Schizosaccharomyces* and *Polysporus* species, and its hydrolysis by a Streptomyces culture filtrate lysing cell walls of *Cryptococcus, Biochim. Biophys. Acta,* 158, 313, 1968.
131. **Matile, P., Cortat, M., Wiemken, A., and Frey-Wyssling, A.,** Isolation of glucanase containing particles from budding *Saccharomyces cerevisiae, Proc. Natl. Acad. Sci. U.S.A.*, 68, 636, 1971.
132. **Phaff, H. T.,** Enzymatic yeast cell wall degradation, *Adv. Chem. Ser.*, 160, 1977.
133. **Farkas, V., Biely, P., and Bauer, S.,** Extracellular β-glucanases of the yeast *Saccharomyces cerevisiae, Biochim. Biophys. Acta,* 321, 246, 1973.
134. **Villa, T. G., Notario, V., and Villanueva, J. R.,** β-Glucanases of the yeast *Pichia polymorpha, Arch. Microbiol.* 104, 201, 1975.
135. **Brock, T. D.** Physiology of the conjugation process in the yeast *Hansenula wingei, J., Gen. Microbiol.*, 26, 487, 1961.
136. **Ahearn, D. G., Meyers, S. P., and Nicols, R. A.,** Extracellular proteinases of yeasts and yeastlike fungi, *Appl. Microbiol.* 16, 1370, 1968
137. **Chen, A. W. and Miller, J. J.,** Proteolytic activity of intact yeast cells during sporulation, *Can. J. Microbiol.*, 14, 957, 1968.
138. **Boulton, A. A.,** Some observations on the chemistry and morphology of the membranes released from yeast protoplasts by osmotic shock, *Exp. Cell Res.*, 37, 343, 1965.
139. **Suomalainen, H., Nurminen, T., and Oura, E.,** Isolation of the plasma membrane of yeast, *Suom. Kemistil.*, B40, 323, 1967.
140. **Dube, J., Setterfield, G., Kiss, G., and Lusena, C. V.,** Fate of the plasma membrane of *Saccharomyces cerevisiae* during rupture, *Can. J. Microbiol.*, 19, 285, 1973.
141. **Willsky, G. R.,** Characterization of the plasma membrane Mg^{2+}-ATPase from the yeast *Saccharomyces cerevisiae, J. Biol. Chem.*, 254, 3326, 1979.

142. **Cabib, E., Ulane, R., and Bowers, B.,** Yeast chitin synthetase. Separation of the zymogen from its activating factor and recovery of the latter in the vacuole fraction, *J. Biol. Chem.*, 248, 1451, 1973.
143. **Scarborough, G. A.,** Isolation and characterization of *Neurospora crassa* plasma membranes, *J. Biol. Chem.*, 250, 1106, 1975.
144. **Duran, A., Bowers, B., and Cabib, E.,** Chitin synthetase zymogen is attached to the yeast plasma membrane, *Proc. Natl. Acad. Sci. U.S.A.*, 72, 3952, 1975.
145. **Abrams, A.,** Reversible metabolic swelling of bacterial protoplasts, *J. Biol. Chem.*, 234, 383, 1959.
146. **Indge, K. J.,** Metabolic lysis of yeast protoplasts, *J. Gen. Microbiol.*, 51, 433, 1968.
147. **Marriott, M. S.,** Isolation and chemical characterization of plasma membranes from the yeast and mycelial forms of *Candida albicans, J. Gen. Microbiol.*, 86, 115, 1975.
148. **Braun, P. C. and Calderone, R. A.,** Regulation and solubilization of *Candida albicans* chitin synthetase, *J. Bacteriol.*, 140, 666, 1979.
149. **Schneider, H., Fuhrmann, G. F., and Fiechter, A.,** Plasma membrane from *Candida tropicalis* grown on glucose or hexadecane. II. Biochemical properties and substrate-induced alterations, *Biochim. Biophys. Acta,* 554, 309, 1979.
150. **Schibeci, A., Rattray, J. B. M., and Kidby, D. K.,** Isolation and identification of yeast plasma membrane, *Biochim. Biophys. Acta,* 311, 15, 1973.
151. **Santos, E., Villanueva, J. R., and Sentandreu, R.,** The plasma membrane of *Saccharomyces cerevisiae.* Isolation and some properties, *Biochim. Biophys. Acta,* 508, 39, 1978.
152. **Keller, F. A. and Cabib, E.,** Chitin and yeast budding. Properties of chitin synthetase from *Saccharomyces carlsbergensis, J. Biol. Chem.*, 246, 160, 1971.
153. **Duran, A. and Cabib, E.,** Solubilization of and partial purification of yeast chitin synthetase. Confirmation of the zymogenic nature of the enzyme, *J. Biol. Chem.*, 253, 4419, 1978.
154. **Cabib, E.,** The yeast primary septum: a journey into three-dimensional biochemistry, *Trends Biochem. Sci.*, 1, 275, 1976.
155. **Betz, H., Hinze, H., and Holzer, H.,** Isolation and properties of two inhibitors of proteinase B from yeast, *J. Biol. Chem.*, 249, 4515, 1974.
156. **Hasilik, A.,** Inactivation of chitin synthase in *Saccharomyces cerevisiae, Arch. Microbiol.*, 101, 295, 1974.
157. **Wolf, D. H. and Ehmann, C.,** Isolation of yeast mutants lacking proteinase B activity, *FEBS Lett.*, 92, 121, 1978.
158. **Zubenko, G. S., Mitchell, A. P. and Jones, E. W.,** Septum formation, cell division, and sporulation in mutants of yeast deficient in proteinase B, *Proc. Natl. Acad. Sci. U.S.A.*, 76, 2395, 1979.
159. **Cabib, E. and Farkas, V.,** The control of morphogenesis: an enzymatic mechanism for the initiation of septum formation in yeast, *Proc. Natl. Acad. Sci. U.S.A.*, 68, 2052, 1971.
160. **Schekman, R. and Brawley, V.,** Localized deposition of chitin on the yeast cell surface in response to mating pheromone, *Proc. Natl. Acad. Sci. U.S.A.*, 76, 645, 1979.
161. **Ruiz-Herrera, J. and Bartnicki-Garcia, S.,** Synthesis of cell wall microfibrils *in vitro* by a "soluble" chitin synthetase from *Mucor rouxii, Science,* 186, 357, 1974.
162. **Lardy, H., Reed, P., and Lin, C. C.,** Antibiotic inhibitors of mitochondrial ATP synthesis, *Fed. Proc. Fed. Am. Soc. Exp. Biol.*, 34, 1707, 1975.
162a. **Willsky,** private communication.
163. **Delhez, J., Dufour, J. P., Thines, D., and Goffeau, A.,** Comparison of the properties of plasma membrane-bound and mitochondrial-bound ATPases in the yeast *Schizosaccharomyces pombe, Eur. J. Biochem.*, 79, 319, 1977.
164. **Dufour, J. and Goffeau, A.,** Solubilization by lysolecithin and purification of the plasma membrane ATPase of the yeast *Schizosaccharomyces pombe, J. Biol. Chem.*, 253, 7026, 1978.
165. **Nurminen, T. and Suomalainen, H.,** The lipolytic activities of the isolated cell envelope fractions of bakers' yeast, *Biochem. J.*, 118, 759, 1970.
166. **Kokke, R., Hooghwinkel, G. J. M., Boij, H. L., van den Bosch, H., Zelles, L., Milder, E., and van Deenan, L. L. M.,** Metabolism of lysolecithin and lecithin in a yeast suspension, *Biochim. Biophys. Acta,* 70, 351, 1963.
167. **Sentandreu, R., Elorza, M. V., and Villaneuva, J. R.,** Synthesis of yeast wall glucan, *J. Gen. Microbiol.* 94, 13, 1975.
168. **Baliant, Š., Farkaš, V., and Bauer, Š.,** Biosynthesis of β-glucans catalyzed by a particulate enzyme preparation from yeast, *FEBS Lett.*, 64, 44, 1976.
169. **Lopez-Romero, E. and Ruiz-Herrera, J.,** Biosynthesis of β-glucans by cell-free extracts from *Saccharomyces cerevisiae, Biochim. Biophys. Acta,* 500, 372, 1977.
170. **Shematek, E. M., Braatz, J. A., and Cabib, E.,** Biosynthesis of the yeast cell wall. I. Preparation and properties of β-(1 $\rightarrow$ 3) glucan synthetase, *J. Biol. Chem.*, 255, 888, 1980.

171. **Shematek, E. M. and Cabib, E.**, Biosynthesis of the yeast cell wall. II. Regulation of β-(1 → 3) glucan synthetase by ATP and GTP, *J. Biol. Chem.*, 255, 895, 1980.
172. **Gunja-Smith, Z. and Smith, E. E.**, Evidence for the periplasmic location of glycogen in *Saccharomyces, Biochem. Biophys. Res. Commun.*, 56, 588, 1974.
173. **Gunja-Smith, Z., Patil, N. B., and Smith, E. E.**, Two pools of glycogen in *Saccharomyces, J. Bacteriol.* 130, 818, 1977.
174. **Wheeler, G. E., Schibeci, A., Epand, R. M., Rattray, J. B., and Kidby, D. K.**, Subcellular localization and some properties of the adenyl cyclase activity of the yeast *Saccharomyces cerevisiae, Biochim. Biophys. Acta,* 372, 15, 1974.
175. **Londesborough, J. C. and Nurminen, T.**, A manganese-dependent adenyl cyclase in baker's yeast *Saccharomyces cerevisiae, Acta Chem. Scand.*, 26, 3396, 1972.
176. **Varimo, K. and Londesborough, J.**, Solubilization and other studies on adenylate cyclase of baker's yeast, *Biochem. J.*, 159, 363, 1976.
177. **Dunlop, P. C. and Roon, R. J.**, L-Asparaginase of *Saccharomyces cerevisiae:* an extracellular enzyme, *J. Bacteriol.*, 122, 1017, 1975.
178. **Dunlop, P. C., Meyer, G. M., Ban, D., and Roon, R. J.**, Characterization of two forms of asparaginase in *Saccharomyces cerevisiae, J. Biol. Chem.*, 253, 1297, 1978.
179. **Dunlop, P. C., Roon, R. J., and Even, H. L.**, Utilization of D-asparagine by *Saccharomyces cerevisiae, J. Bacteriol.*, 125, 999, 1976.
180. **Kaplan, J. G.**, Action of non-penetrating heavy metals on the catalase activity of yeast cells, *Nature (London),* 205, 76, 1965.
181. **Kaplan, J. G.**, Heat inactivation of cryptic catalase of yeast, *Nature (London),* 196, 950, 1962.
182. **Singh, B. and Datta, A.**, Induction of *N*-acetylglucosamine-catabolic pathway in spheroplasts of *Candida albicans, Biochem. J.*, 178, 427, 1979.
183. **Snyder, H. E. and Phaff, H. J.**, Studies on a β-fructosidase (inulinase) produced by *Saccharomyces fragilis, Antonie van Leeuwenhoek; J. Microbiol. Serol.*, 26, 433, 1960.
184. **Hopkins, R. H.**, Amylase systems in culture and wild yeasts, *Wallerstein Lab. Commun.*, 21, 309, 1958.
185. **Andrews, J. and Gilliland, R. B.**, Super-attenuation of beer: a study of three organisms capable of causing abnormal attenuations, *J. Inst. Brew. (London),* 58, 189, 1952.
186. **Wickerham, L. J., Lockwood, L. B., Pettijohn, O. G., and Ward, G. E.**, Starch hydrolysis and fermentation by the yeast *Endomycopsis fibuligera, J. Bacteriol.*, 48, 413, 1944.
187. **Villa, T. G., Notario, V., and Villaneuva, J. R.**, Occurrence of an endo-1, 3-β-glucanase in culture fluids of the yeast *Candida utilis, Biochem. J.*, 177, 107, 1979.
188. **Villa, T. G., Notario, V., Benitz, T., and Villanueva, J. R.**, Purification of an exo-1, 3-β-glucanase from *Candida utilis, Can. J. Biochem.*, 54, 927, 1976.
189. **Meyer, M. T. and Phaff, H. J.**, Survey for α-(1 → 3)-glucanase activity among yeasts, *J. Bacteriol.*, 131, 702, 1977.
190. **Abd-El-Al, A. T. and Phaff, H. J.**, Purification and properties of endo-β-glucanase in the yeast *Hanseniaspora valbyensis, Can. J. Microbiol.*, 15, 697, 1969.
191. **Abd-El-Al, A. T. and Phaff, H. J.**, Exo-β-glucanases in yeast, *Biochem. J.*, 109, 347, 1968.
192. **Ahearn, D. G., Meyers, S. P., and Nichols, R. A.**, Extracellular proteinases of yeasts and yeastlike fungi, *Appl. Microbiol.*, 16, 1370, 1968.
193. **Maddox, I. S. and Hough, J. S.**, Proteolytic enzymes of *Saccharomyces carlsbergensis, Biochem. J.*, 117, 843, 1970.
194. **Dodyk, F. and Rothstein, A.**, Factors influencing the appearance of invertase in *Saccharomyces cerevisiae, Arch. Biochem. Biophys.*, 104, 478, 1964.
195. **Elorza, M. V., Villanueva, J. R., and Sentandreu, R.**, The mechanism of catabolite inhibition of invertase by glucose in *Saccharomyces cerevisiae, Biochim. Biophys. Acta,* 475, 103, 1977.
196. **Holbein, B. E., Forsberg, C. W., and Kidby, D. K.**, A modified procedure for studying enzyme secretion in yeast sphaeroplasts: subcellular distribution of invertase, *Can. J. Microbiol.*, 22, 989, 1976.
197. **Davies, R.**, Enzyme formation in *Saccharomyces fragilis.* I. Invertase and raffinase, *Biochem. J.*, 55, 484, 1953.
198. **Weimberg, R.**, Repression of the acid phosphatase of *Saccharomyces bisporus* in relation to the polyphosphate content of the cells, *Can. J. Microbiol.*, 22, 867, 1976.
199. **Davies, A.**, Invertase formation in *Saccharomyces fragilis, J. Gen. Microbiol.*, 14, 109, 1956.
200. **Palade, G. E., Siekevitz, P., and Caro, L. G.**, in *The Exocrine Pancreas,* DeReuck, A. V. S. and Cameron, M. P., Eds., Churchill Livingstone, London, 1962, 23.
201. **Tokuyasu, K. and Sherbaum, O. H.**, Ultrastructure of mucocysts and pellicle of *Tetrahymena pyriformis, J. Cell. Biol.*, 27, 67, 1965.

202. **Satir, B., Schooley, C., and Satir, P.,** Membrane reorganization during secretion in *Tetrahymena, Nature (London),* 235, 53, 1972.
203. **Moor, H.,** Endoplasmic reticulum as the initiator of bud formation in yeast *(S. cerevisiae), Arch. Mikrobiol.,* 57, 135, 1967.
204. **Nickerson, W. J.,** Molecular basis of form in yeast, *Bacteriol. Rev.,* 27, 305, 1963.
205. **Novick, P. and Schekman, R.,** Secretion and cell-surface growth are blocked in a temperature-sensitive mutant of *Saccharomyces cerevisiae, Proc. Natl. Acad. Sci. U.S.A.,* 76, 1858, 1979.
206. **Blobel, G. and Dobberstein, B.,** Transfer of proteins across membranes. I. Presence of proteolytically processed and unprocessed nascent immunoglobulin light chains on membrane bound ribosomes of murine myeloma, *J. Cell. Biol.,* 67, 835, 1975.
207. **Sanders, R. L. and May, B. K.,** Evidence for extrusion of unfolded extracellular enzyme polypeptide chains through membranes of *Bacillus amyloliquefaciens, J. Bacteriol.,* 123, 806, 1975.
208. **Lazdunski, C., Baty, D., and Pages, J. M.,** Procaine, a local anesthetic interacting with the cell membrane, inhibits the processing of precursor forms of periplasmic proteins in *Escherichia coli, Eur. J. Biochem.,* 96, 49, 1979.
209. **Inouye, H. and Beckwith, J.,** *Proc. Natl. Acad. Sci. U.S.A.,* 74, 1438, 1977.
210. **Holbein, B. E. and Kidby, D. K.,** Effects of proteolytic enzymes on invertase secretion in sphaeroplasts of *Saccharomyces:* inhibition by trypsin, *Can. J. Microbiol.,* 23, 202, 1977.
211. **Arnold, W. N., Garrison, R. G., and Boyd, K. S.,** Periplasmic structure in *Saccharomyces rouxii* (Boutroux), an osmophil, *Appl. Microbiol.,* 28, 1047, 1974.
212. **Moore, R. T.,** The ultrastructure of fungal cells, in *The Fungi,* Vol. 1, Ainsworth, G. S. and Sussman, A. S., Eds., Academic Press, New York, 1965, 95.
213. **Ghosh, B. K., Sargent, M. G., and Lampen, J. O.,** Morphological phenomena associated with penicillinase induction and secretion in *Bacillus licheniformis, J. Bacteriol.,* 96, 1314, 1968.
214. **Nordlie, R. C.,** Multifunctional glucose-6-phosphatase: cellular biology, *Life Sci.,* 24, 2397, 1979.

Chapter 2

BIOSYNTHETIC MECHANISMS FOR CELL ENVELOPE POLYSACCHARIDES

A. J. Parodi

TABLE OF CONTENTS

I. INTRODUCTION

This chapter will deal with the pathways leading to the synthesis of the primary polysaccharides of yeast cell envelopes, namely glucan, chitin, and mannan, as well as with the localization of the glycosyltransferases and some of the mechanisms by which the formations of the polymers are thought to be regulated. Results obtained with other fungal systems and also with animal tissues will be mentioned, especially when knowledge on yeast systems is scarce or when information gathered using other organisms can help in the understanding of the synthetic process. Further information on the location of some of the key synthetic enzymes in yeast cell envelopes is to be found in Volume II, Chapter 1. The disposition of polymers within envelopes is addressed in Volume I, Chapter 5.

II. GLUCAN

About 85% of the glucan in *Saccharomyces cerevisiae* is composed of a β (1-3)-linked polymer containing about 3% of β (1-6) bonds.[1] There are also two other less prevalent glucans that contain higher proportions (19 and 10%) of β (1-6) linkages.[2,3] Although these polymers are responsible for the mechanical strength of the cell wall (see Volume I, Chapter 3) and play major roles in the determination of shape and the preservation of integrity of the cell, many intriguing questions remain about their mechanism of biosynthesis. The reason for this is that the preparation of cell-free extracts containing appreciable amounts of glucan synthetase activity only recently has become practical.

A. Glucan Biosynthesis in Yeasts

A glucan synthetase activity that required UDP-[^{14}C]Glc as a substrate was detected by Sentandreu et al.[4] in *Sacch. cerevisiae* cells that had been made permeable with a mixture of toluene and ethanol. The incorporation of [^{14}C] glucose from this precursor was enhanced in the presence of Mg^{++} ions. Subcellular fractionation showed that most of the label was present in a fraction sedimenting between 10,000 × g and 40,000 × g. No enzymatic activity, however, could be demonstrated in the isolated particles. The authors postulated that glucan might be synthesized inside cytoplasmic vesicles that then fuse with the plasma membrane and subsequently translocate the polysaccharide to the cell wall proper. More recent evidence does not support this hypothesis. As will be commented on below, the drastic method employed[4] to break the cells, i.e., bombardment of cells with glass beads, has profound effects on membranous structures and can lead to misleading results when the subcellular distribution of enzymes or polysaccharides is being investigated.

The first report on glucan synthesis by *Sacch. cerevisiae* cell-free preparations was that of Bálint et al.[5] A membrane fraction sedimenting between 10,000 × g and 100,000 × g, also obtained from cells disrupted by mechanical procedures, contained two independent enzymatic activities that used GDP-Glc and UDP-Glc as their respective substrates. The products in each case had β (1-3)- and β (1-6)-linked glucose units; however, the polymer formed by the UDP-Glc-specific enzymatic activity had a higher proportion of β (1-3) bonds. Both enzymes were stimulated by Mn^{++} ions, but only in the case of the glucan synthetase that recognized UDP-Glc could Mn^{++} be replaced by Mg^{++} ions. Interestingly, the product formed from GDP-Glc was partially susceptible to β-elimination in mild alkali, thus suggesting that it might be bound to protein through serine or threonine residues. It will be seen below that some of the properties of the glucan synthetases described by Bálint et al.[5], such as the metal requirement or the recognition of GDP-Glc as substrate, are in disagreement with later studies. As the specific activities reported[5] were extremely low, the possibility exists that the glucans whose synthesis they described represent only minor components of the family of similar polysaccharides. Glucan synthetase activities were found by López-Romero and Ruiz-Herrera[6] both in membrane- and cell wall-rich fractions prepared (as in the previous cases) from *Sacch. cerevisiae* cells disrupted by mechanical disintegration. The former fraction had a higher total activity but lower specific activity than the latter. Both particulate fractions contained enzymes that catalyzed the addition of β (1-3)- and β (1-6)-linked glucose residues but the proportion of the latter activity was higher in the cell wall-rich preparation. Both enzymatic activities were specific for UDP-Glc. Only trace amounts of glucose units were incorporated when GDP-Glc was substituted for UDP-Glc. An interesting finding was that both the specific and total activities of β (1-3) glucan synthetase diminished as the cultures passed from mid- to late-logarithmic and finally into stationary phase.

Shematek et al.[7] found only a β (1-3) glucan synthetase activity in a membrane fraction derived from *Sacch. cerevisiae* protoplasts. This is the most efficient enzymatic preparation reported to date. Its specific activity was about 80-fold higher than that reported by López-Romero and Ruiz-Herrera[6] and the synthesis of the polysaccharide in assays supplemented with several activators (see below) proceeded at about 20% of the in vivo rate.[7] There was no evidence for the involvement of lipid intermediates.[7] That is, no radioactive compounds could be detected in organic-solvent extracts of reaction mixtures incubated with UDP-[^{14}C]Glc, and the reaction was not enhanced by the addition of dolichol-P. Moreover, dolichol-P-[^{14}C] glucose could not be used instead of UDP-[^{14}C]Glc and the synthesis of glucan proceeded better with EDTA than with Mg^{++} ions. (In all systems studied so far, bivalent cations such as Mg^{++} or Mn^{++} are required for the synthesis of Dol-P-Glc.[8] As will be seen in Section IV.A., dolichol-P-Glc is an intermediate in the synthesis of some yeast glycoproteins.) The glucose transfer in the synthesis of glucan was specific for UDP-[^{14}C]Glc.[7] ADP[^{14}C]Glc or GDP-[^{14}C]Glc were ineffective. Equivalent amounts of UDP were formed for each glucose residue transferred to the polymer. The glucan product contained only β (1-3)-linked glucose units and had a chain length of 60 to 80. The equation for β (1-3) glucan synthesis is

$$\text{UDP-Glc} + [\beta(1\text{-}3)\text{Glc}]_n \rightarrow [\beta(1\text{-}3)\text{Glc}]_{n+1} + \text{UDP}$$

Apparently the synthesis of glucan proceeded *de novo* since labeled sorbitol residues were detected after reduction of the isolated polymer followed by total acid hydrolysis.[7] This result does not rule out the possibility that an acceptor other than glucose could serve as primer for the initiation of the polymerization. For instance, it is known that saccharide moieties joined to serine or threonine residues in proteins are liberated under the conditions employed by Shematek et al.[7] for the reduction of the reaction product with $NaBH_4$. The above-mentioned results were obtained with mixed membrane preparations; however, all of the enzymatic activity was found in a fraction containing almost pure plasma membranes and the enzyme is said to be disposed towards the cytoplasmic side. The methodology employed here was similar to that used in the case of chitin synthetase (see also Chapter 1 in this volume). A trypsin treatment similar to the one (see below) used to change the zymogen form of chitin synthetase into the active enzyme completely supressed the β (1-3) glucan synthetase activity. Other relevant features described by Schematek et al.[7] were the six- to eightfold activation of β (1-3) glucan synthetase by the individual addition of bovine serum albumin, glycerol, GTP, or ATP. The effect of bovine serum albumin was not due simply to the addition of protein because ovalbumin was ineffective. Moreover, serum albumin did not protect the enzyme against inactivation at 30°C. These results suggest that perhaps the serum protein could be mimicking an intracellular, more specific macromolecular activator lost during the purification of the enzyme preparation.

B. Evidence for a Regulatory Mechanism

During the budding cycle of *Sacch. cerevisiae,* new glucan is deposited exclusively in the daughter cells. Synthesis ceases once mother and daughter cells are separated and is reinitiated only upon the emergence of new buds[9] (see also Volume II, Chapter 3). Extant studies indicate that β (1-3) glucan synthetase is intimately associated with the plasma membrane, and because there is neither hindrance to the flow of cytoplasmic material between bud and mother cell nor discontinuity in the plasma membrane that surrounds them, it is difficult to envisage a simple chemical mechanism by which only part of the glucan synthetase molecules are activated or inhibited. The effect of bovine serum albumin, the activation by nucleoside triphosphates, and the lower activity present in stationary phase cells mentioned above may be reflections of a regulatory mechanism.

The effects of nucleoside triphosphates on glucan synthetase have been studied in some detail by Shematek and Cabib.[10] They propose a hypothetical model in which it is assumed that ATP phosphorylates (or adenylates) a substance loosely attached to the particulate preparation. This substance, designated P_1X in this scheme, is lost upon centrifugation and can be dephosphorylated by *Escherichia coli* alkaline phosphatase. P_1X can be further modified at the expense of ATP to P_1XP_2. This compound, or GTP, can activate β (1-3) glucan synthetase once it is bound to the enzyme. Both activators may occupy the same site.

E. coli alkaline phosphatase or similar endogenous activities that can be inhibited by EDTA, but activated by Mg^{++} ions may dephosphorylate P_1XP_2 thus inactivating glucan synthetase. Although this model is simply a working hypothesis the existing data at least implies that yeast β (1-3) glucan synthetase is regulated by a sophisticated mechanism.

C. Glucan Biosynthesis in Other Fungi and Comparison with Yeast Systems

Because there is a paucity of reports on glucan synthesis by yeast cell-free preparations, results obtained with other fungal systems will be mentioned as they may help to suggest working hypotheses for the regulation of yeast glucan synthesis. According to Nanba and Kuroda[11,12] cell homogenates of the fungus *Cochliobolus mayabeanus* were able to catalyze the transference of monosaccharide residues from UDP-Glc to a polymer. The reaction product contained both β (1-3)- and β (1-6)-linked glucose units. A cell wall-rich fraction from *Neurospora crassa* was reported to contain a β (1-3) glucan synthetase that recognized UDP-Glc as a substrate.[13] Results on glucan synthesis by particulate preparations from the plant pathogenic fungus *Phytophthora cinnamomi* have certain similarities to those obtained with yeast systems; thus fractions that contained cell wall fragments[14] had both β (1-3) and β (1-6) glucan synthetases whereas a pure membranous preparation[15] was devoid of the latter activity. The sugar nucleotide recognized by both synthetases was UDP-Glc. As mentioned earlier, López-Romero and Ruiz-Herrera[6] found that in *Sacch. cerevisiae* a higher proportion of the β (1-6) synthetase was in a cell-wall-containing fraction and Schematek et al.[7] were unable to detect this activity in a pure plasma membrane preparation. It is possible that the locations of the β (1-3) and β (1-6) glucosyl transferases in yeasts, the former in the plasma membrane and the latter apparently somehow associated with the cell wall, may determine the disposition of the major glucans.[16]

The addition of boiled supernat to the *P. cinnamomi* particulate preparations enhanced the incorporation of glucose residues into glucan about eightfold.[14] It is not known whether this represents an activation analogous to that reported for *Sacch. cerevisiase*.[10] In contradistinction to the latter system, incubation of the *P. cinnamomi* membranous preparations with trypsin resulted in a nearly twofold enhancement of glucan synthesis, thus suggesting the presence of the enzyme in a zymogen state.[15] Alternatively, the trypsin treatment may have rendered an active but somehow masked enzyme available to the substrates.

The synthesis of glucans from UDP-Glc by particulate fractions prepared from mycelia of the oomycete *Saprolegnia monoica* has been described by Fèvre and Dumas.[17] In this case the reaction product was mainly a cellulose-like material containing β (1-4) linkages. The synthesis of the polysaccharide was stimulated by cellobiose and $MgCl_2$. Little activity was present in the supernatant fraction following high speed centrifugation. The supernatant fraction contained, however, a heat-labile inhibitor of the particulate enzyme. Most of the glucan synthetase activity was present in a cell-wall-rich fraction, but activity was also detected in membrane fractions apparently derived from the Golgi apparatus, the endoplasmic reticulum, and the plasma membrane; however, since drastic methods were used in order to disrupt the cells, no far-reaching conclusions can be drawn from these results.

III. CHITIN

Although chitin, a β (1-4)-linked polymer of *N*-acetylglucosamine, is one of the major components of the cell wall in filamentous fungi, in yeasts it may be restricted to the bud scars. The budding process including the mechanism by which the chitin-containing septum is formed will be dealt with in Volume II, Chapter 3. Consequently, only the chemical features of the biosynthesis of chitin will be mentioned here.

A. Sugar Donor and Primer Requirement

The immediate sugar donor in chitin synthesis has been found to be UDP-GlcNAc. No evidence for the existence of a lipid intermediate could be found with partially purified preparations of *Sacch. cerevisiae* chitin synthetase solubilized with digitonin.[18] Dolichol-P did not stimulate the transference of glycosyl residues to chitin nor could labeled substances soluble in organic solvents be isolated when UDP-[^{14}C]GlcNAc of high specific activity was employed. Moreover, tunicamycin was without effect at concentrations (5 μg/m*l*) at which it greatly inhibited the synthesis of dolichol-P-P-GlcNAc in yeast systems.[19] Chitin synthetase appeared to be about 20% inhibited at 100 μg/m*l* of antibiotic.[18] A similarly small inhibition by high concentrations of tunicamycin has been reported for *N. crassa* chitin synthetase.[20] This effect is not surprising since the antibiotic has a *N*-acetylglucosamine residue in its molecule and is, therefore, a competitive inhibitor for UDP-GlcNAc. The general equation for chitin synthesis can be written as follows:

$$\text{UDP-GlcNAc} + [\beta(1\text{-}4)\text{GlcNAc}]_n \rightarrow [\beta(1\text{-}4)\text{GlcNAc}]_{n+1} + \text{UDP}$$

The above-mentioned soluble, partially purified chitin synthetase preparation, was shown to be able to catalyze the synthesis of milligram quantities of the polymer.[18] Although these results argue against the necessity for a macromolecular primer, they do not preclude the possibility that the enzymatic preparations could have had tightly bound oligosaccharides as acceptors. For instance, homogenous *E. coli* glycogen synthetase preparations were found to contain oligosaccharides noncovalently attached to the enzyme.[21]

B. Localization of Chitin Synthetase

Disruption of cells by violent procedures causes rupture and vesiculation of the plasma and other membranes and results in the production of variously contaminated fragments.[22] This may explain several claims that fungal chitin synthetases are located in the cell wall.[13,23,24] Durán et al.[25] circumvented this problem in their study of chitin synthetase from *Sacch. cerevisiae* by using a technique originally devised by Scarborough[26] for the isolation of *N. crassa* plasma membranes. Apparently the plasma membrane is obtained in larger unsealed fragments if it previously has been coated with concanavalin A. Omission of this pretreatment with the lectin results in an extensive fragmentation of the membrane that hampers the separation of this organelle from intracellular membranes. Concanavalin A-coated plasma membranes were purified in Renografin gradients and found to contain nearly all of the chitin synthetase activity.[25] The enzyme appeared to be disposed towards the inner cytoplasmic side of the plasma membrane as shown by various sequential treatments involving potential inactivation of chitin synthetase by glutaraldehyde before and after lysis.[25] Similar results were offered by Braun and Calderone[27] for both yeast and hyphal forms of *Candida albicans.*

In the fungus *Phycomyces blakesleeanus,* subcellular fractionation showed that most of the chitin synthetase co-sedimented with a fraction containing 5′-nucleotidase activity which is a good marker for plasma membranes in eukaryotic cells.[28] In contrast, it has been reported that in yeast-form cells of *Mucor rouxii,* chitin synthetase is associated with rather large particles.[29] The latter had a 35 to 100 nm diameter, a molecular weight higher than 7×10^6, and an average sedimentation coefficient of 105 S. The size of these so-called chitosomes suggests that they are not intrinsic constituents of membranes. The latter has an average thickness of only 7 to 9 nm. It is not known if the chitosomes do really exist in vivo or if they are artifacts created by the drastic procedure employed for cell disruption. The latter possibility is suggested by the fact that enzymatic complexes totally unrelated to chitin synthesis, such as fatty acid synthetase and pyruvate dehydrogenase, were found to be present in the chitosome preparations.[29,30]

An intermediate situation between *Sacch. cerevisiae* and *M. rouxii* has been described for *Schizophyllum commune.*[31] The distribution of chitin synthetase was studied in subcellular fractions prepared from lysed concanavalin A-coated protoplasts. About half of the total enzymatic activity was associated with the membrane, whereas 30% sedimented between $10{,}000 \times g$ and $300{,}000 \times g$. The membrane-associated enzyme was largely in an active state, whereas the enzyme present in the other fractions could be activated about 40-fold by a proteolytic treatment. As studies performed using electron microscopy and high resolution autoradiography of regenerating *S. commune* protoplasts had indicated that chitin synthesis was associated with the plasma membrane, the authors speculated that the intracellular inactive form of chitin synthetase could represent the enzyme being conveyed from its site of synthesis to the plasma membrane.[31]

C. Regulation of Chitin Synthetase Activity

Although chitin synthetase in yeasts seems to be located throughout the plasma membrane, the synthesis of chitin occurs only at the same time and place as the formation of the septum between bud and mother cell.[25,32] This implies that chitin synthetase activity must be regulated by a stringent mechanism. As first described for *Sacch. cerevisiae* and later for other fungi, particulate preparations of chitin synthetase can be activated by a limited treatment with fungal proteases or with similar enzymes from other sources such as papain, trypsin, or chymotrypsin.[27,31-40] In the yeast case, the endogenous activating protease is perhaps proteinase B, an enzyme present in the vacuolar fraction.[33,41] Other yeast proteases are ineffective.[35] The possibility existed, however, that the membrane-bound chitin synthetase always could be in an active form but trapped in such a way that substrates were inaccessible to it. According to this hypothesis the activation by proteases would be due to an unmasking rather than to a proteolytic cleavage of the enzyme. The fact that soluble, partially purified preparations of chitin synthetase from *Sacch. cerevisiae* were also found to be activated by trypsin argues against this possibility.[18] A heat-stable protein present in the yeast cytoplasm has been shown to bind to proteinase B and thus prevent the activation of chitin synthetase.[32,42] In the case of *M. rouxii,* the inhibitor did not interact with the proteolytic activator but reacted directly with the activated chitin synthetase.[43]

Based on the results obtained with *Sacch. cerevisiae* Farkaš and Cabib[32] proposed a mechanism for the regulation of chitin synthesis. According to this model the activator (proteinase B) would transform the zymogen (inactive) form of chitin synthetase into an active state upon fusion of the proteinase B-containing vacuole with the plasma membrane at the site of septum formation. The cytoplasmic inhibitor would prevent the action of activator molecules unspecifically leaked from the vacuole. There is no proof, however, that this mechanism operates in vivo. In fact, *Sacch. cerevisiae* mutants devoid of proteinase B were shown to grow at the same rate and to have both the same morphology and insertion of chitin in the septum as the wild type cells.[44,45] It is possible that a less prevalent yet unidentified yeast protease could be the physiological activator of chitin synthetase. In connection with these findings it should be mentioned that although chitin synthetase isolated from both yeast and hyphal forms of *C. albicans* appeared to be largely in an inactive state, a factor capable of activating the enzyme only could be isolated from the yeast-like cells.[27] Another possibility is that the in vitro proteolytic activation of chitin synthetase could be grossly mimicking another mechanism of regulation. For instance, glycogen synthetase from *Sacch. cerevisiae* can exist in two forms, one independent and the other dependent upon glucose-6-phosphate for activity.[46] The dependent form is phosphorylated, whereas the independent one is not. The passage from the dependent to the independent or more active form can be brought about in vitro either by enzymatic removal of phosphate residues or by a proteolytic treatment that liberates a phosphate-containing peptide.[46] The former is the physiological mechanism for glycogen synthetase activation.

IV. MANNAN

Mannan is a generic name for different polysaccharides having in common a high proportion of mannose residues. Unlike chitin or glucan, mannans are readily soluble in water once extracted from the cells and do not appear to contribute to the shape of the organisms. They are usually covalently bound to protein, and apparently their role in yeasts is to hold together the different components of the yeast cell wall (see Volume I, Chapter 5). The best studied structure is that of *Sacch. cerevisiae* mannan.[47-50] As depicted in Volume I, Chapter 5, Figure 1, mannan has two types of saccharide moieties. One type is composed of an inner core containing 12 to 17 mannose and two *N*-acetylglucosamine units joined to an asparagine residue of the protein and then an outer chain formed by 100 to 150 mannoses. The two differences between the inner core and the outer chain regions are (1) the existence of phosphomannose residues in the outer chain, and (2) the presence in the inner core of mannose units joined with α (1-3) bonds to the α (1-6)-linked polymannose backbone. The second type of saccharide moiety in *Sacch. cerevisiae* mannan is composed of mannose, mannobiose, mannotriose, and mannotetraose residues linked to serine/threonine units in the protein. It must be mentioned that the evidence showing that both asparagine– and serine/threonine-linked saccharides are joined to the same protein is not conclusive. One yeast mannoprotein is known, the specific intercellular agglutination factor from *Hansenula wingei,* that contains only the latter type of saccharide moiety.[51]

A. Biosynthesis of Asparagine-Linked Saccharides

Most studies on the role of derivatives of dolichol as intermediates in the glycosylation of proteins of the asparagine-linkage type have been performed with animal systems.[8] For this reason, occasional mention of results obtained with higher organisms will help in understanding the pathway. Dolichol is a polyprenol containing 17 to 21 and 14 to 18 isoprene units in animal tissues[52] and yeasts,[53] respectively.

The first step in the glycosylation process (Figure 1) is the synthesis of dolichol-P-P-GlcNAc upon reaction of dolichol-P and UDP-GlcNAc.[54,55] Another *N*-acetyl-glucosamine and up to 9 mannose and three glucose residues then may be added to dolichol-P-P-GlcNAc.[52,53,56-61] It has been found, at least for animal tissues, that the precursors of the second *N*-acetylglucosamine and the first mannose residue are in each case the respective sugar nucleotides, whereas the substrate for the transfer of most of the rest of the mannose units is dolichol-P-Man.[8] In both animal tissues and yeasts, the donor of the glucose residues is dolichol-P-Glc.[8,61] Dolichol-P-Glc and dolichol-P-Man are synthesized through the reaction of dolichol-P and UDP-Glc or GDP-Man, respectively. Incubation of yeast particulate preparations with dolichol-P-P-$(GlcNAc)_2$, or the same plus one or more mannose residues, or with the derivative having in addition glucose units, leads to the transference of the sugar moieties to endogenous proteins.[53,56-61] The real precursor of glycoproteins in animal tissues appeared to be the dolichol-P-P derivative containing the full complement of monosaccharides, that is, two *N*-acetylglucosamine, nine mannose, and three glucose residues.[62-64] This compound is the main derivative existing in vivo. It has been found that in the synthesis of the "complex" type animal glycoproteins, i.e., those containing two *N*-acetylglucosamines, three mannoses, and distal monosaccharides as *N*-acetylglucosamines, galactoses, sialic acids, and fucoses, the transference of the oligosaccharide from the lipid derivative to the protein is followed by the trimming of the glucose and most of the mannose residues. The distal monosaccharides are then added directly from the respective sugar nucleotides. Only the glucose and fewer mannose residues than in the previous case are apparently trimmed in the synthesis of the "simple" type glycoproteins which have only two *N*-acetylglucosamine and more than three mannose units.

Yeast cells have the "simple" type of glycoproteins and seem to lack the "complex" one; however, as will be seen below, the fact that mannose residues may be added directly from GDP-Man to oligosaccharides already bound to proteins as in mannan, confers on these glycoproteins a formal resemblance to the "complex" ones at least from a biosynthetic point of view. *Sacch. cerevisiae* strain ATCC 9763 cells appeared to have roughly similar amounts of four dolichol-P-P derivatives containing, respectively, two *N*-acetylglucosamine and probably nine mannose and the same plus one, two, or three glucose residues.[65] Partial characterization of the saccharide moieties suggested that they have a similar structure to those in animal tissues.[65] However, there are differences in the mannose linkages between the structures reported for the animal cell lipid-linked oligosaccharide[64] and the inner core of yeast mannan.[47]

The pattern of oligosaccharides previously bound to asparagine residues isolated from *Sacch. cerevisiae* cells pulsed for a very short time with [U-^{14}C] glucose was very similar to that of the saccharides linked to dolichol-P-P. It is not known, however, if the four protein-bound oligosaccharides found after the pulse had been transferred from the respective dolichol-P-P derivatives, or if, for instance, only the largest one came from the lipid derivative, and hence, the pattern found was due to the trimming of some

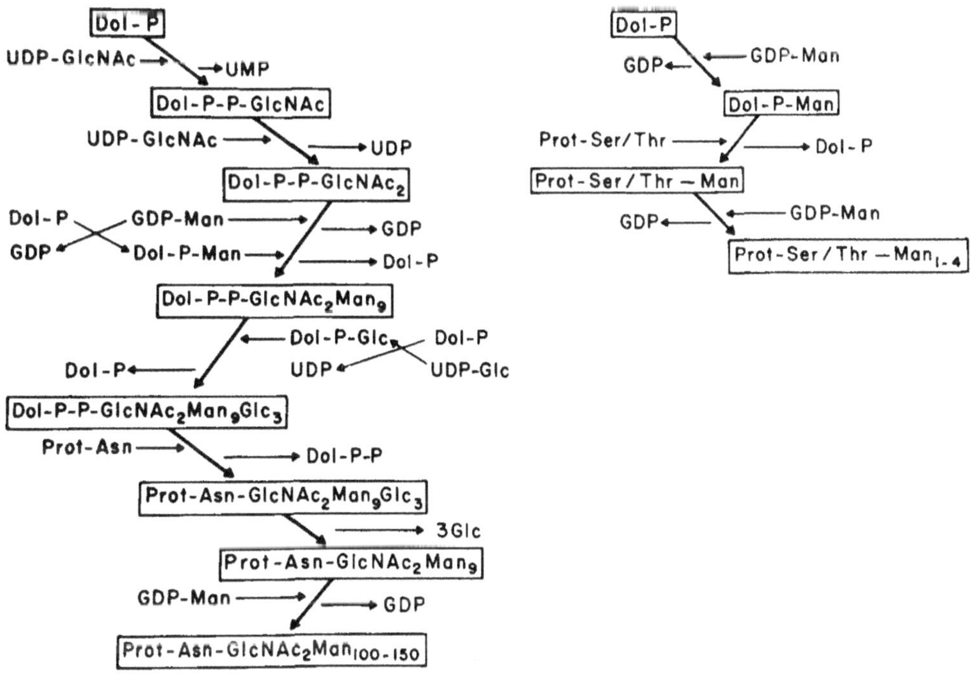

FIGURE 1. Schematic representation of the pathways leading to the synthesis of the saccharide moieties of *Saccharomyces cerevisiae* mannoprotein. The following abbreviations are employed: Dol = dolichol, P = phosphate, UDP = uridine diphosphate, GlcNAc = *N*-acetylglucosamine, UMP = uridine monophosphate, GDP = guanosine diphosphate, Man = mannose, Glc = glucose, Prot = protein, Asn = asparagine, Ser = serine, and Thr = threonine.

monosaccharides from the protein-bound oligosaccharide. The latter possibility seems to be true in other strains of *Sacch. cerevisiae* in which the largest oligosaccharide has been found to be virtually the only one joined to dolichol-P-P in vivo.[65a] Samples taken after short chase periods showed that the protein-bound oligosaccharides gradually lost their glucose residues and were converted into the compound containing two *N*-acetylglucosamine and nine mannose residues.[65] On longer chase periods, however, the latter oligosaccharide was either trimmed or enlarged by further addition of mannose residues, probably depending on the glycoprotein being synthesized.[65] The transference of mannoses to oligosaccharides already bound to proteins could represent, at least in part, the completion of the inner core, as the latter may contain from 12 to 17 mannose residues and apparently only 9 of them at the most are transferred from the dolichol-P-P derivative.[47,65] Results presented by Nakajima and Ballou[47] suggest that the completion of the inner core is brought about by the same enzymes involved in the synthesis of mannan outer chain. Thus mutants devoid of the terminal α (1-3)-linked mannose residues in the side branches of the outer chain also lacked those residues in the inner core. As will be seen in the next section, the sugar donor for the formation of the outer chain is GDP-Man and the mannosyltransferases involved in these reactions are different from those responsible for the synthesis of dolichol-P-P-oligosaccharides.[66]

B. Biosynthesis of Mannan Outer Chain

Early work of Cabib and co-workers[67,68] showed that mannose residues in the polysaccharide part of yeast mannan originated from GDP-Man; however, as dolichol derivatives were unknown at that time and the studies on mannan synthesis had been performed with particulate preparations that may have contained endogenous dolichol-P, the question of whether lipid intermediates were involved in the transfer reactions was not clarified until recently. Thus Lehle and Tanner reported that incubation of GDP-[^{14}C]Man with *Sacch. cerevisiae* particulate preparations led to the incorporation of radioactivity into both asparagine- and serine/threonine-linked saccharides, but only the latter became labeled when dolichol-P-[^{14}C]Man was used as a substrate.[57] The absence of lipid intermediates in the synthesis of the polysaccharide part of mannan was confirmed when it was reported that the enzymes involved in the formation of the outer chain could be separated from those responsible for the synthesis of dolichol-P-Man and dolichol-P-P-oligosaccharides.[66] This was accomplished by ion-exchange column chromatography of a soluble enzymatic extract prepared by treating a yeast particulate fraction with urea and detergents. These results also showed that both sets of mannosyltransferases were different. Supporting this notion it was found that the in vitro synthesis of the outer chain required the presence of Mn^{++} ions, and the respective enzymes were stable upon storage of the soluble preparation at 2°C, whereas the enzymes involved in the synthesis of dolichol-P-Man and dolichol-P-P-oligosaccharides were active in the presence of Mg^{++} or Mn^{++} ions and were rapidly inactivated when kept at low temperature.[66] Apparently the mannose residues present in different positions in the outer chain are added by distinct mannosyltransferases, and the structure of the acceptor determines which enzyme will be operative. These results were found using short mannooligosaccharides as substrates. For instance, a simulation of the extension of the α (1-6)-linked polymannose backbone can be achieved with the saccharide αMan(1-3)αMan(1-2)αMan(1-2)Man which serves as an acceptor for only an α(1-6)-linked mannose residue at the reducing end if *Sacch. cerevisiae* enzymes are used;[69] however, extracts prepared from *Sacch. chevalieri* catalyze also the transference of a mannose unit to position three at the nonreducing terminus.[69] It should be noted that the latter yeast species (unlike *Sacch. cerevisiae*) may contain a pentasaccharide in the side branches of the outer chain. An α(1-6)-linked mannotetraose which can be considered as a model of the backbone accepts mannoses bound to it with α(1-6) and α(1-2) bonds, thus mimicking both the extension of the backbone and the first step in the formation of the side branches.[66,69] It has been found that the K_m of the α(1-2)- and α(1-6)-mannosyltransferases for α(1-6)-linked mannooligosaccharides decreases from the disaccharide to the tetrasaccharide and then increases with higher degrees of polymerization of the acceptor substrate.[69] This kinetic property of the enzymes undoubtedly has a decisive influence on the degree of branching of mannan.

The synthesis of the phosphomannose residues has been studied in *H. holstii*,[70,71] *H. capsulata*[72] and *Sacch. cerevisiae*[73] extracts. In all three species it has been found that the substrate for this reaction is GDP-Man, but in this case mannose-1-phosphate and not the monosaccharide alone is transferred. Other fungal species may have residues different from mannose in their outer chains. In some cases studies on their mechanism of biosynthesis have been reported. For example, the transference of *N*-acetylglucosamine residues from UDP-GlcNAc to macromolecular endogenous acceptors or to exogenous mannooligosaccharide acceptors is catalyzed by *Kluyveromyces lactis* particulate preparations.[74] The fungus *Penicillium charlesii* has a cell wall glycopeptide containing both a phosphogalactomannan polysaccharide and short mannooligosaccharides, the latter being bound to serine/threonine residues.[75,76] GDP-Man is the donor for the residues in the polysaccharide. The in vitro transference requires Mn^{++} ions, and no lipid intermediates seem to be involved.

Several yeast mutants affected in the structure of their mannan outer chains have been isolated.[50] In some cases the structural alteration could be correlated with a specific enzymatic defect. Thus, the lack of mannotetraoses in the side chains of the mannan extracted from *Sacch. cerevisiae* mnn_1 mutant was found to be due to the absence of the mannosyltransferase involved in the attachment of the α(1-3)-linked terminal residues.[66,69] This and other cases in which such a correlation between structural and enzymatic defects could be established support the notion of the presence of a distinct mannosyltransferase for the mannoses present in different positions in the outer chain. Furthermore, the fact that mutants having alterations in their outer chains synthesize the same dolichol-P-P-oligosaccharides as the wild type strains is also evidence for the dissimilarity between the mannosyltransferases responsible for the formation of the lipid derivatives and those involved in the synthesis of the outer chain.[66] In other cases, however, no correlation between structural and enzymatic defects could be made. Thus, the outer chain of *Sacch. cerevisiae* mnn_2 mutants was composed of only the α(1-6)-linked polymannose backbone, but the cells had normal levels of the α(1-2) mannosyltransferase responsible for the first step in the synthesis of the side branches.[66] Similarly, the mannan of *K. lactis* $mnn_{2\text{-}2}$ mutant lacked *N*-acetylglucosamine residues in its outer chain, but the same level of the respective transferase was found in the mutant and in the wild type strains.[74]

C. Biosynthesis of Serine/Threonine-Linked Saccharides

Dolichol-P-Man is an intermediate in the transference of the residues attached to serine/threonine whereas the rest of the mannoses are added directly from GDP-Man (Figure 1). It was found that incubation of dolichol-P-[^{14}C]Man with *Sacch. cerevisiae* particulate preparations led to the mannosylation of endogenous proteins.[77,78] The label could be released as mannose if the reaction product was subjected to a mild alkaline treatment that liberates serine/threonine- but not asparagine-linked saccharides. Labeled mannobiose and mannotriose in addition to mannose were released by the alkaline treatment if unlabeled GDP-Man was added to the reaction mixtures containing dolichol-P-[^{14}C]Man, but the radioactivity was exclusively located at the reducing ends. It should be mentioned that up to now no evidence for the involvement of dolichol derivatives in the synthesis of serine/threonine-linked animal glycoproteins has been presented. Apparently the same enzyme is responsible for the addition of the α(1-3)-linked terminal mannose residues in *Sacch. cerevisiae* outer chains and serine/threonine-linked moieties because the same mutants are devoid of such residues in both locations.[48]

It has been reported that dolichol-P-Man is also an intermediate in the glycosylation of serine/threonine residues in other fungi, such as *H. holstii*, *N. crassa*, *Fusarium solani*, and *Aspergillus niger*.[79-82] The evidence in the first three cases was based on direct transfer from dolichol-P-Man, whereas in the latter report only kinetic results were provided. One of the components of the cell wall of *Cryptococcus laurentii* is a glycoprotein containing both a pentasaccharide and a dodecasaccharide joined to serine/threonine residues in the protein moiety. The former saccharide is composed of 4 mannose and 1 xylose units and the dodecasaccharide contains 11 galactose and 1 mannose residues. The residue bound to the amino acids is mannose in both cases. Studies on the in vitro synthesis of these oligosaccharides indicated that most if not all the monosaccharides were transferred by different glycosyltransferases.[83,84] No clear involvement of lipid intermediates in the reactions was shown. The in vitro transference of mannose residues from GDP-Man to the serine/threonine-bound mannooligosaccharides of *P. charlesii* cell wall glycopeptide has been reported.[75,76] The transference required Mn^{++} ions and apparently no lipid derivatives acted as intermediates. This is not surprising since analysis of the reaction products showed that the mannoses had been transferred to saccharides already bound to protein and not to the serine/threonine residues.

D. Localization of Mannosyltransferases

Early reports showed that mannan, unlike glucan or chitin, was synthesized in the cytosol. Thus, Košinová et al.[85] pulse-chased yeast cells with minute amounts of radioactive mannose in the presence of large quantities of glucose. Under these conditions all the label is incorporated into mannan. Autoradiographic analysis of thin sections of the cells showed that the radioactivity was initially present in the cytoplasm and appeared afterwards in the cell wall. Cortat et al.[86] separated different subcellular fractions in Urografin gradients and found that mannan synthesis was highest in a fraction containing internal membranes. These reports, however, did not distinguish between mannose residues transferred directly from GDP-Man or through lipid intermediates, nor did they clearly identify the organelles where the transference of the sugar residues took place.

It has been found that in animal cells the oligosaccharide is transferred from the dolichol-P-P derivative to incompleted polypeptide chains. As the protein is being synthesized, it passes from the cytoplasm to the lumen of the rough endoplasmic reticulum where the saccharide moiety is added.[87-89] According to the location of the specific glycosidases and glycosyltransferases involved in the process, trimming of the glucose and mannose residues from the glycoproteins begins in the endoplasmic reticulum and ends in the Golgi apparatus, whereas the addition of the distal monosaccharides in the synthesis of the "complex" glycoproteins occurs in the Golgi apparatus.[90,91]

There are indications that in fungi the transference of the oligosaccharide to asparagine and of mannose to serine/threonine residues also occurs at the polysomal level. Ruiz-Herrera and Sentandreu pulsed *Sacch. cerevisiae* cells with [^{14}C]mannose, stopped protein synthesis with cycloheximide, and isolated the polysomes.[92] The latter was found to contain radioactivity that could be released at least in part upon incubation of the crude polysomal preparation with puromycin, thus indicating that the label was bound to nascent polypeptide chains. Total acid hydrolysis and paper chromatography of the labeled material present in the polysomes revealed that the radioactivity was in mannose, glucose, and glucosamine residues.[92] They probably came from the hydrolysis of the oligosaccharide transferred from the dolichol-P-P derivative to protein, since, as mentioned in Section IV.A., that saccharide is composed of the same monosaccharide units.

Long and Rudick[93] also found radioactivity exclusively associated with the polysomes extracted from *A. niger* cells pulsed with [^{14}C]mannose. Preincubation of the cells with tunicamycin resulted in a 80% decrease in the incorporation of label. Since this antibiotic inhibits the synthesis of dolichol-P-P-GlcNAc, the first step in the formation of dolichol-P-P-oligosaccharides, it can be concluded that the label associated with the polysomes had been transferred to incompleted polypeptide chains through dolichol-P-P derivatives. Glycosylation of serine/threonine residues appeared to occur, at least in part, at the polysomal level.

Larriba et al.[94] incubated a cycloheximide-containing lysate from *Sacch. cerevisiae* protoplasts with dolichol-P-[^{14}C]Man and found that part of the radioactivity was present in the polysomal fraction. The label could be released as mannose upon a mild alkaline treatment. Results obtained by Marriott and Tanner[95] with subcellular fractions from sucrose gradients of a lysate from *Sacch. cerevisiae* protoplasts, indicated that both the total and specific activities for the synthesis of dolichol-P-Man, dolichol-P-P-GlcNAc, and dolichol-P-P-(Glc NAc)$_2$, as well as for the transference of sugar residues from dolichol-P-Man and GDP-Man to mild akali-labile and resistant reaction products, were higher in the rough endoplasmic reticulum. Somewhat lower enzymatic levels were found in the smooth endoplasmic reticulum. Previous work using Urografin gradients reached similar conclusions, although in this case no distinction between rough and smooth endoplasmic reticula was made.[96] The last two studies do not decide if the synthesis of the entire saccharide moieties of yeast mannoproteins occurs in the endoplasmic reticulum, or if for instance, part of it takes place in the so-called glucanase-rich vesicles. The latter is said to contain substantial amounts of the activities involved in the formation of outer chains and in the addition of the distal residues to the serine/threonine-linked saccharides.[96] It has been suggested that the glucanase-rich vesicles are responsible for the completion of glycosylation and for the transport of the mannoproteins to the cell exterior as they fuse with the plasma membrane.[96]

E. Regulation of Mannan Synthesis

The participation of more than 20 different glycosyltranferases in the pathway leading to mannan synthesis probably precludes the existence of a regulatory mechanism affecting all the enzymes in a similar way. Apparently mannan synthesis is regulated solely by the availability of new protein acceptors for the transferases. Inhibition of protein synthesis with cycloheximide stops mannan synthesis in protoplasts or intact cells although such treatment may even stimulate chitin or glucan synthesis.[97,98] The presence of that antibiotic does not affect the in vitro activity of mannosyltransferases.[98] Moreover, no significant difference between the specific activities of mannosyltransferases involved in the synthesis of the outer chain isolated from logarithmic- and stationary-phase cells was found.[65a]

All these results suggest, therefore, that as protein synthesis proceeds the polysaccharide is formed by mannosyltransferases constitutively present at internal membranes; however, the new mannan is deposited by an unknown mechanism, exclusively in the buds.[99] The structure of mannan is apparently determined exclusively by the stringent specificity of the mannosyltransferases towards the acceptors (Section IV.B.).

As expected, tunicamycin inhibits the synthesis of mannan but not that of glucan or protein.[100] The fate of the protein moiety of mannan in the presence of the antibiotic is unknown. Information on this subject can be obtained, however, from yeast enzymes that have mannooligosaccharides of different sizes attached to them. Addition of tunicamycin to yeast cultures may result in three different responses. In the case of alkaline phosphatase, which normally contains a small amount of carbohydrate, the accumulation of the enzyme is not affected, although as expected the carbohydrate-free form of the protein is produced.[101] With β-fructofuranosidase and acid phosphatase the accumulation of enzymes is totally abolished.[100] An intermediate situation occurs with yeast carboxypeptidase Y whose accumulation is diminished in the presence of the antibiotic.[102] These dissimilar results may be due to the fact that the absence of the sugar residues modifies the conformation of the polypeptide chains. This in turn may affect proteins in different ways, as shown by results obtained with carbohydrate-free animal and yeast glycoproteins synthesized in the presence of tunicamycin. Some proteins, such as fibronectin, become more susceptible to proteolytic attack.[103] Other proteins, such as immunoglobulins, have their migration through cell inner membranes hampered or present a reduced solubility (e.g., vesicular stomatitis virus G protein).[104,105] Finally, some carbohydrate-free proteins, such as yeast alkaline phosphatase, may not be affected at all.[101] Early proposals suggesting a feedback mechanism between glycosylation and protein synthesis seem to be ruled out.[102]

V. CONCLUDING REMARKS

Advances in knowledge of the pathways of polysaccharide and glycoprotein synthesis have been impressive in the last few years; however, there are many questions remaining about fundamental aspects of cell envelopes. Regulation is necessary because the formation of these polysaccharides has both temporal and spatial restraints. Presently there are only hints about the ways in which syntheses of chitin and glucan are regulated. Furthermore, although these polysaccharides are mostly located in the cell wall, extant evidence indicates that the respective synthetases are bound to the plasma membrane and at least parts of the active sites face the cytoplasm. This suggests that a translocation or extrusion of the growing polysaccharide chains must occur across the plasma membrane. How this translocation is brought about, and even what is actually translocated, is unknown. Similarly, the passage of mannan from the vicinity of internal membranes (where it seems to be synthesized) to the cell wall is just beginning to be studied.

Even some primary chemical processes remain obscure. For instance, why are lipid derivatives only involved in the transfer of those residues that are proximal to the protein moiety of mannan? Concerning glucan, it is not known if the branching points are constituted by transferases acting on nucleoside diphosphate glucose donors or whether a branching enzyme breaks an existing (1,3) linkage and forms a new (1,6) linkage. The latter possibility is best illustrated in the synthesis of glycogen. Also, the possible involvement of lipid derivatives in the formation of glucan branching points can not yet be excluded.

The main studies on the biosynthesis of the polysaccharides of the yeast cell envelope have been performed with *Sacch. cerevisiae* as a model system. Some polysaccharides present in other yeasts have very different structures but in most cases no studies on the mechanisms of their biosyntheses have been performed. For instance, the yeasts *Schizosaccharomyces pombe* and *Schiz. versatilis* contain an α(1-3)-bound glucan,[106] but to date no reports on an α(1-3) glucan synthetase have been published.

There is a paucity of literature on the biosynthesis of envelope components by pathogenic yeast and yeast-like organisms. These are eukaryotic cells and the existence of a cell wall (and, in some cases, a slime layer) is an obvious but fundamental distinction from the cells of the higher eukaryotic host-animal. Therefore, the steps leading to the synthesis of cell wall components are logical targets for inhibition of yeast cell division and hence for the potential development of specific antifungal drugs.

REFERENCES

1. **Manners, D. J., Masson, A. J., and Patterson, J. C.,** The structure of a β-(1-3)-D-glucan from yeast cell walls, *Biochem. J.*, 135, 19, 1973.
2. **Manners, D. J., Masson, A. J., Patterson, J. C., Björndal, H., and Lindberg, B.,** The structure of a β-(1,6)-D-glucan from yeast cell walls, *Biochem. J.*, 135, 31, 1973.
3. **Fleet, G. H. and Manners, D. J.,** Isolation and composition of an alkali-soluble glucan from the cell wall of *Saccharomyces cerevisiae, J. Gen. Microbiol.*, 94, 180, 1976.
4. **Sentandreu, R., Elorza, M. V., and Villanueva, J. R.,** Synthesis of yeast wall glucan, *J. Gen. Microbiol.*, 94, 13, 1975.
5. **Bálint, Š., Farkaš, V., and Bauer, Š.,** Biosynthesis of β-glucans catalyzed by a particulate enzyme preparation from yeast, *FEBS Lett.*, 64, 44, 1976.
6. **López-Romero, E. and Ruiz-Herrera, J.,** Biosynthesis of β-glucans by cell-free extracts from *Saccharomyces cerevisiae, Biochim. Biophys. Acta*, 500, 372, 1977.
7. **Shematek, E. M., Braatz, J. A., and Cabib, E.,** Biosynthesis of the yeast cell wall. I. Preparation and properties of a β(1-3)glucan synthetase, *J. Biol. Chem.*, 255, 888, 1980.
8. **Parodi, A. J. and Leloir, L. F.,** The role of lipid intermediates in the glycosylation of proteins in the eucaryotic cell, *Biochim. Biophys. Acta*, 559, 1, 1979.

9. **Biely, P., Kovařík, J., and Bauer, Š.**, Cell wall formation in yeast. An electron microscopic autoradiographic study, *Arch. Mikrobiol.*, 94, 365, 1973.
10. **Shematek, E. M. and Cabib, E.**, Biosynthesis of the yeast cell wall. II. Regulation of β(1-3)glucan synthetase by ATP and GTP, *J. Biol. Chem.*, 255, 895, 1980.
11. **Nanba, H. and Kuroda, H.**, Studies on fungicides. X. Biosynthesis of β-glucan and chitin-like substance of cell wall from *Cochliobolus miyabeanus, Chem. Pharm. Bull.*, 22, 610, 1974.
12. **Nanba, H. and Kuroda, H.**, Studies on fungicides. XI. Biosynthesis of β-glucan and chitin-like substance of cell wall from *Cochliobolus miyabeanus, Chem. Pharm. Bull.*, 22, 1895, 1974.
13. **Mishra, N. C. and Tatum, E. L.**, Effect of sorbose on polysaccharide synthetases of *Neurospora crassa, Proc. Natl. Acad. Sci. U.S.A.*, 69, 313, 1972.
14. **Wang, M. C. and Bartnicki-Garcia, S.**, Biosynthesis of β-1,3- and β-1,6-linked glucan by *Phytophthora cinnamomi* hyphal walls, *Biochem. Biophys. Res. Commun.*, 24, 832, 1966.
15. **Wang, M. C. and Bartnicki-Garcia, S.**, Synthesis of β-1,3-glucan microfibrils by a cell-free extract from *Phytophhora cinnamomi, Arch. Biochem. Biphys.*, 175, 351, 1976.
16. **Kopecká, M., Phaff, H. J., and Fleet, G. H.**, Demonstration of a fibrillar component in the cell wall of the yeast *Saccharomyces cerevisiae* and its chemical nature, *J. Cell. Biol.*, 62, 66, 1974.
17. **Fèvre, M. and Dumas, C.**, β-Glucan synthetases from *Saprolegnia monoica, J. Gen. Microbiol.*, 103, 297, 1977.
18. **Durán, A. and Cabib, E.**, Solubilization and partial purification of yeast chitin synthetase. Confirmation of the zymogenic nature of the enzyme, *J. Biol. Chem.*, 253, 4419, 1978.
19. **Lehle, L. and Tanner, W.**, The specific site of tunicamycin inhibition in the formation of dolichol-bound *N*-acetylglucosamine derivatives, *FEBS Lett.*, 71, 167, 1976.
20. **Selitrennikoff, C. P.**, Competitive inhibition of *Neurospora crassa* chitin synthetase activity by tunicamycin, *Arch. Biochem. Biophys.*, 195, 243, 1979.
21. **Holmes, E. and Preiss, J.**, Characterization of *Escherichia coli B* glycogen synthase enzymatic reactions and products, *Arch. Biochem. Biophys.*, 196, 436, 1979.
22. **Dubé, J., Setterfield, G., Kiss, G., and Lusena, C. V.**, Fate of plasma membrane of *Saccharomyces cerevisiae* during cell rupture, *Can. J. Microbiol.*, 19, 285, 1973.
23. **McMurrough, I., Flores-Carreón, A., and Bartnicki-Garcia, S.**, Pathway of chitin synthesis and cellular localization of chitin synthetase in *Mucor rouxii, J. Biol. Chem.*, 246, 3999, 1971.
24. **Peberdy, J. F. and Moore, P. M.**, Chitin synthetase in *Mortirella vinacea,* properties, cellular location and synthesis in growing cultures, *J. Gen. Microbiol.*, 90, 228, 1975.
25. **Durán, A., Bowers, B., and Cabib, E.**, Chitin synthetase zymogen is attached to the yeast plasma membrane, *Proc. Natl. Acad. Sci. U.S.A.*, 72, 3952, 1975.
26. **Scarborough, G. A.**, Isolation and characterization of *Neurospora crassa* plasma membranes, *J. Biol. Chem.*, 250, 1106, 1975.
27. **Braun, P. C. and Calderone, R. A.**, Chitin synthesis in *Candida albicans:* comparison of yeast and hyphal forms, *J. Bacteriol.*, 133, 1472, 1978.
28. **Jan, Y. N.**, Properties and cellular localization of chitin synthetase in *Phycomyces blakesleeanus, J. Biol. Chem.*, 249, 1973, 1974.
29. **Ruiz-Herrera, J., López-Romero, E., and Bartnicki-Garcia, S.**, Properties of chitin synthetase in isolated chitosomes from yeast cells of *Mucor rouxii, J. Biol. Chem.*, 252, 3338, 1977.
30. **Ruiz-Herrera, J., Sing, V. O., van der Voude, V. J., and Bartnicki-Garcia, S.**, Microfibril assembly by granules of chitin synthetase, *Proc. Natl. Acad. Sci. U.S.A.*, 72, 2706, 1975.
31. **Vermeulen, C. A., Raeven, M. B. J. M., and Wessels, J. G. H.**, Localization of chitin synthetase activity in subcellular fractions of *Schizophyllum commune* protoplasts, *J. Gen. Microbiol.*, 114, 87, 1979.
32. **Cabib, E. and Farkaš, V.**, The control of morphogenesis: an enzymatic mechanism for the initiation of septum formation in yeast, *Proc. Natl. Acad. Sci. U.S.A.*, 68, 2052, 1971.
33. **Archer, D. B.**, Chitin biosynthesis in protoplasts and subcellular fractions of *Aspergillus fumigatus, Biochem. J.*, 164, 653, 1977.
34. **Gooday, G. W. and De Rousset-Hall, A.**, Properties of chitin synthetase from *Coprinus cinereus, J. Gen. Microbiol.*, 89, 137, 1975.
35. **Hasilik, A.**, Inactivation of chitin synthase in *Saccharomyces cerevisiae, Arch. Microbiol.*, 101, 295, 1974.
36. **López-Romero, E. and Ruiz-Herrera, J.**, Synthesis of chitin by particulate preparations from *Aspergillus flavus, Antonie van Leeuwenhoek; J. Microbiol. Serol.*, 42, 261, 1976.
37. **Ruiz-Herrera, J. and Bartnicki-Garcia, S.**, Synthesis of cell wall microfibrils *in vitro* by a "soluble" chitin synthetase from *Mucor rouxii, Science,* 186, 357, 1974.
38. **Ryder, N. S. and Peberdy, J. F.**, Chitin synthetase in *Aspergillus nidulans:* properties and proteolytic activation, *J. Gen. Microbiol.*, 99, 69, 1977.

39. **Van Laere, A. J. and Carlier, A. R.,** Synthesis and proteolytic activation of chitin synthetase in *Phycomyces blakesleeanus* Burgeff, *Arch. Microbiol.*, 116, 181, 1978.
40. **Arroyo-Begovich, A., and Ruiz-Herrera, J.,** Proteolytic activation and inactivation of chitin synthase from *Neurospora crassa, J. Gen. Microbiol.*, 113, 339, 1979.
41. **Hasilik, A. and Holzer, H.,** Participation of the tryptophan synthase inactivating system from yeast in the activation of chitin synthase, *Biochem. Biophys. Res. Commun.*, 53, 552, 1973.
42. **Ulane, R. E. and Cabib, E.,** The activating system of chitin synthetase from *Saccharomyces cerevisiae.* Purification and properties of an inhibitor of the activating factor, *J. Biol. Chem.*, 249, 3418, 1974.
43. **López-Romero, E., Bartnicki-Garcia, S., and Ruiz-Herrera, J.,** Purification of inhibitory protein of chitin synthetase from *Mucor rouxii, Biochim. Biophys. Acta,* 525, 338, 1978.
44. **Zubenko, G. S., Mitchell, A. P., and Jones, E. W.,** Septum formation, cell division and sporulation in mutants of yeast deficient in proteinase B, *Proc. Natl. Acad. Sci. U.S.A.*, 76, 2395, 1979.
45. **Wolf, D. H. and Ehmann, C.,** Studies on a proteinase B mutant of yeast, *Eur. J. Biochem.*, 98, 375, 1979.
46. **Huang, K. P. and Cabib, E.,** Yeast glycogen synthetase in the glucose-6-phosphate-dependent form. II. The effects of proteolysis, *J. Biol. Chem.*, 249, 3859, 1974.
47. **Nakajima, T. and Ballou, C. E.,** Microheterogeneity of the inner core region of yeast mannoprotein, *Biochem. Biophys. Res. Commun.*, 66, 870, 1975.
48. **Nakajima, T. and Ballou, C. E.,** Characterization of the carbohydrate fragments obtained from *Saccharomyces cerevisiae* mannan by alkaline degradation, *J. Biol. Chem.*, 249, 7679, 1974.
49. **Nakajima, T. and Ballou, C. E.,** Structure of the linkage region between the polysaccharide and protein parts of *Saccharomyces cerevisiae* mannan, *J. Biol. Chem.*, 249, 7685, 1974.
50. **Ballou, C. E.,** Some aspects of the structure, immunochemistry and genetic control of yeast mannans, *Adv. Enzymol. Relat. Areas Mol. Biol.*, 40, 239, 1974.
51. **Yen, P. H. and Ballou, C. E.,** Composition of a specific intercellular agglutination factor, *J. Biol. Chem.*, 248, 8316, 1973.
52. **Dunphy, P. J., Kerr, J. D., Pennock, J. F., Whittle, K. J., and Fenney, J.,** The plurality of long chain isoprenoid alcohols (polyprenols) from natural sources, *Biochim. Biophys. Acta,* 136, 136, 1967.
53. **Jung, P. and Tanner, W.,** Identification of the lipid intermediate in yeast mannan biosynthesis, *Eur. J. Biochem.*, 37, 1, 1973.
54. **Lehle, L. and Tanner, W.,** Formation of lipid-bound oligosaccharides in yeast, *Biochim. Biophys. Acta,* 399, 364, 1975.
55. **Reuvers, F., Habets-Willems, C., Reinking, A., and Boer, P.,** Glycolipid intermediates involved in the transfer of *N*-acetylglucosamine to endogenous proteins in a yeast membrane preparation, *Biochim. Biophys. Acta,* 486, 541, 1977.
56. **Nakayama, K., Araki, V., and Ito, E.,** The formation of a mannose-containing trisaccharide on a lipid and its transfer to proteins in yeast, *FEBS Lett.*, 72, 287, 1976.
57. **Lehle, L. and Tanner, W.,** Glycosyl transfer from dolichyl-phosphate sugars to endogenous and exogenous glycoprotein acceptors in yeast, *Eur. J. Biochem.*, 83, 563, 1978.
58. **Parodi, A. J.,** Lipid intermediates in the synthesis of the inner core of yeast mannan, *Eur. J. Biochem.*, 83, 253, 1978.
59. **Lehle, L. and Tanner, W.,** Biosynthesis and characterization of large dolichylphosphate-linked oligosaccharides in *Saccharomyces cerevisiae, Biochim. Biophys. Acta,* 539, 218, 1978.
60. **Parodi, A. J.,** Protein glycosylation through dolichol derivatives in baker's yeast, *FEBS Lett.*, 71, 283, 1976.
61. **Parodi, A. J.,** Synthesis of glucosyl-dolichol derivatives in yeast and their role in protein glycosylation, *Eur. J. Biochem.*, 75, 171, 1977.
62. **Tabas, I., Schlesinger, S., and Kornfeld, S.,** Processing of high mannose oligosaccharides to form complex type oligosaccharides on the newly synthesized polypeptides of the vesicular stomatitis virus G protein and the IgG heavy chain, *J. Biol. Chem.*, 253, 716, 1978.
63. **Robbins, P. W., Hubbard, S. C., Turco, S. J., and Wirth, D. F.,** Proposal for a common oligosaccharide intermediate in the synthesis of membrane glycoproteins, *Cell,* 12, 893, 1977.
63a. **Hunt, L. A., Etchison, J. R., and Summers, D. F.,** Oligosaccharide chains are trimmed during synthesis of the envelope glycoprotein of vesicular stomatitis virus, *Proc. Natl. Acad. Sci. U.S.A.*, 75, 754, 1978.
64. **Li, E., Tabas, I., and Kornfeld, S.,** The synthesis of complex-type oligosaccharides. I. Structure of the lipid-linked oligosaccharide precursor of the complex-type oligosaccharides of the vesicular stomatitis virus G protein, *J. Biol. Chem.*, 253, 7762, 1978.
65. **Parodi, A. J.,** Biosynthesis of yeast glycoproteins. Processing of the oligosaccharides transferred from dolichol derivatives, *J. Biol. Chem.*, 254, 10051, 1979.
65a. **Parodi, A. J.,** Unpublished data.

66. **Parodi, A. J.,** Biosynthesis of yeast mannoproteins. Synthesis of mannan outer chain and of dolichol derivatives, *J. Biol. Chem.*, 254, 8343, 1979.
67. **Algranati, I. D., Carminatti, H., and Cabib, E.,** The enzymic synthesis of yeast mannan, *Biochem. Biophys. Res. Commun.*, 12, 504, 1963.
68. **Behrens, N. H. and Cabib, E.,** The biosynthesis of mannan in *Saccharomyces cerevisiae, J. Biol. Chem.*, 243, 502, 1968.
69. **Nakajima, T. and Ballou, C. E.,** Yeast manno-protein biosynthesis: solubilization and selective assay of four mannosyltransferases, *Proc. Natl. Acad. Sci. U.S.A.*, 72, 3912, 1975.
70. **Bretthauer, R. K., Kozak, C. P., and Irwin, W. E.,** Phosphate and mannose transfer from guanosine diphosphate mannose to yeast mannan acceptors, *Biochem. Biophys. Res. Commun.*, 37, 820, 1969.
71. **Kozak, L. P. and Bretthauer, R. K.,** Studies on the biosynthesis of *Hansenula holstii* mannans from guanosine diphosphate mannose, *Biochemistry,* 9, 1115, 1970.
72. **Mayer, R. M.,** The enzymatic synthesis of the phosphomannan of *Hansenula capsulata, Biochim. Biophys. Acta,* 252, 39, 1971.
73. **Karson, E. and Ballou, C. E.,** Biosynthesis of yeast mannan. Properties of a mannosylphosphate transferase in *Saccharomyces cerevisiae, J. Biol. Chem.*, 253, 6484, 1978.
74. **Smith, W. L., Nakajima, T., and Ballou, C. E.,** Biosynthesis of yeast mannan. Isolation of *Kluyveromyces lactis* mannan mutants and a study of the incorporation of *N*-acetyl-D-glucosamine into the polysaccharide side chains, *J. Biol. Chem.*, 250, 3426, 1975.
75. **Gander, J. E., Drewes, L. R., Fang, F., and Lui, A.,** 5-D-β-D-galactofuranosyl-containing exocellular glycopeptide of *Penicillium charlesii.* Incorporation of mannose from GDP-D-mannose into glycopeptide, *J. Biol. Chem.*, 252, 2187, 1977.
76. **Gander, J. E. and Fang, F.,** Properties of *Penicillium* GDP-mannose: glycopeptide mannosyl transferase solubilized with Triton X-100, *J. Supramol. Struct.*, 6, 579, 1977.
77. **Babczinski, P. and Tanner, W.,** Involvement of dolichol monophosphate in the formation of specific mannosyl linkages in yeast glycoproteins, *Biochem. Biophys. Res. Commun.*, 54, 1119, 1973.
78. **Sharma, C. B., Babczinski, P., Lehle, L., and Tanner, W.,** The role of dolichol monophosphate in glycoprotein biosynthesis in *Saccharomyces cerevisiae, Eur. J. Biochem.*, 46, 35, 1974.
79. **Bretthauer, R. K. and Wu, S.,** Synthesis of the mannosyl-*O*-serine (threonine) linkage of glycoproteins from polyisoprenyl phosphate mannose in yeast *(Hansenula holstii), Arch. Biochem. Biophys.*, 167, 151, 1975.
80. **Gold, M. H. and Hahn, H. J.,** Role of a mannosyl lipid intermediate in the synthesis of a *Neurospora crassa* glycoproteins, *Biochemistry,* 15, 1808, 1976.
81. **Soliday, C. L. and Kolattukudy, P. E.,** Introduction of *O*-glycosidically linked mannose into proteins via mannosyl phosphoryl dolichol by microsomes from *Fusarium solani* f. *pisi, Arch. Biochem. Biophys.*, 197, 367, 1979.
82. **Létoublon, R. and Got, R.,** Rôle d'un intermédiaire dans le transfert du mannose à des accepteurs glycoprotéiques endogènes chez *Aspergillus niger, FEBS Lett.*, 46, 214, 1974.
83. **Schutzbach, J. S. and Ankel, H.,** Multiple mannosyl transferases in *Cryptococcus laurentii, J. Biol. Chem.*, 246, 2187, 1971.
84. **Schutzbach, J. S., Raizada, M. K., and Ankel, H.,** Heteroglycan synthesis in *Cryptococcus laurentii, J. Biol. Chem.*, 249, 2953, 1974.
85. **Košinová, A., Farkaš, V., Machala, S., and Bauer, Š.,** Site of mannan synthesis in yeast. An autoradiographic study, *Arch. Microbiol.*, 99, 255, 1974.
86. **Cortat, M., Matile, P., and Kopp, F.,** Intracellular localization of mannan synthetase activity in budding baker's yeast, *Biochem. Biophys. Res. Commun.*, 55, 482, 1973.
87. **Kiely, M. L., McKnight, G. S., and Schimke, R. T.,** Studies on the attachment of carbohydrate to ovalbumin nascent chains in hen oviduct, *J. Biol. Chem.*, 251, 5490, 1976.
88. **Katz, F. N. and Lodish, H. F.,** Transmembrane biogenesis of the Vesicular Stomatitis virus glycoprotein, *J. Cell. Biol.*, 80, 416, 1979.
89. **Czichi, U. and Lennarz, W. J.,** Localization of the enzyme system for glycosylation of proteins via the lipid-linked pathway in rough endoplasmic reticulum, *J. Biol. Chem.*, 252, 7901, 1977.
90. **Ugalde, R., Staneloni, R., and Leloir, L. F.,** Action of glycosidases on the saccharide moiety of the glucose containing dolichyl diphosphate oligosaccharide, *FEBS. Lett.*, 91, 209, 1978.
91. **Schachter, H., Jabbal, I., Hudgin, R. L., Pinteric, L., McGuire, E. J., and Roseman, S.,** Intracellular localization of liver sugar nucleotide glycoprotein glycosyltransferases in a Golgi-rich fraction, *J. Biol. Chem.*, 245, 1090, 1970.
92. **Ruiz-Herrera, J. and Sentandreu, R.,** Site of initial glycosylation of mannoproteins from *Saccharomyces cerevisiae, J. Bacteriol.*, 124, 127, 1975.
93. **Long, S. D. and Rudick, M. J.,** Glycosylation of nascent polypeptide chains in *Aspergillus niger, Arch. Biochem. Biophys.*, 198, 212, 1979.

94. **Larriba, G., Elorza, M. V. Villanueva, J. R., and Sentandreu, R.,** Participation of dolichol phospho-mannose in the glycosylation of yeast wall manno-proteins at the polysomal level, *FEBS Lett.*, 71, 316, 1976.
95. **Marriott, M. and Tanner, W.,** Localization of dolichyl phosphate- and pyrophosphate-dependent glycosyl transfer reactions in *Saccharomyces cerevisiae, J. Bacteriol.*, 139, 565, 1979.
96. **Lehle, L., Bauer, F., and Tanner, W.,** The formation of glycosidic bonds in yeast glycoproteins. Intracellular localization of the reactions, *Arch. Microbiol.*, 114, 77, 1977.
97. **Elorza, M. V. and Sentandreu, R.,** Effect of cycloheximide on yeast cell wall synthesis, *Biochem. Biophys. Res. Commun.*, 36, 741, 1969.
98. **Farkaš, V., Svoboda, A., and Bauer, Š.,** Secretion of cell wall glycoproteins by yeast protoplasts. The effect of 2-deoxy-D-glucose and cycloheximide, *Biochem. J.*, 118, 755, 1970.
99. **Farkaš, V., Kovařík, J., Košinová, A., and Bauer, Š.,** Autoradiographic study of mannan incorporation into the growing cell walls of *Saccharomyces cerevisiae, J. Bacteriol.*, 117, 265, 1974.
100. **Kuo, S. C. and Lampen, J. O.,** Tunicamycin an inhibitor of yeast glycoprotein synthesis, *Biochem. Biophys. Res. Commun.*, 58, 287, 1974.
101. **Onishi, H. R., Tkacz, J. S., and Lampen, J. O.,** Glycoprotein nature of yeast alkaline phosphatase. Formation of active enzyme in the presence of tunicamycin, *J. Biol. Chem.*, 254, 11943, 1979.
102. **Hasilik, A. and Tanner, W.,** Carbohydrate moiety of carboxypeptidase Y and perturbation of its biosynthesis, *Eur. J. Biochem.*, 91, 567, 1978.
103. **Olden, K., Pratt, R. M., and Yamada, K. M.,** Role of carbohydrates in protein secretion and turnover: effects of tunicamycin on the major cell surface glycoprotein of chick embryo fibroblasts, *Cell,* 13, 461, 1978.
104. **Hickman, S., Kulczycki, A., Jr., Lynch, R. G., and Kornfeld, S.,** Studies of the mechanism of tunicamycin inhibition of IgA and IgE secretion by plasma cells, *J. Biol. Chem.*, 252, 4402, 1977.
105. **Leavitt, R., Schlesinger, S., and Kornfeld, S.,** Impaired intracellular migration and altered solubility of nonglycosylated glycoproteins of vesicular stomatitis virus and Sindbis virus, *J. Biol. Chem.*, 252, 9018, 1977.
106. **Bacon, J. S. D., Jones, D., Farmer, V. C., and Webley, D. M.,** The occurrence of α(1-3)glucan in *Cryptococcus, Schizosaccharomyces* and *Polyporus* species, and its hydrolysis by a *Streptomyces* culture filtrate lysing cell walls of *Cryptococcus, Biochim. Biophys. Acta,* 158, 313, 1968.

Chapter 3

MOLECULAR BIOLOGY OF BUDDING

M. L. Slater

TABLE OF CONTENTS

I. INTRODUCTION

The great majority of yeast species undergo vegetative reproduction by budding. In this chapter *Saccharomyces cerevisiae* is taken as the model because of the preponderance of data on this species. A perusal of the micrographs of Volume I, Chapter 2, will give some idea of the number of untouched morphological problems associated with more exotic architecture.

The molecular biology of budding is inherently interesting in terms of the underlying mechanisms which attend three dimensional changes in the cell envelope. Moreover, the yeast cell provides some unique features of eukaryotic cell division. For example, the great dissimilarity in mass between mother and separated daughter cell (under modest growth rates) has no parallel in organisms that divide by binary fission.[1,2]

Elucidation of budding mechanisms can be approached with a variety of techniques which are primarily biochemical and genetic in scope. However, one should emphasize at the outset that the bud is identifiable by light microscopy and no other procedure is required in this regard. The bud is recognized by its shape rather than its composition, and that shape is intimately associated with the structure of the cell wall.[3] The composition of the wall can explain its ability to *maintain* cell-shape, and consequently the distinction between the bud and the rest of the cell. But how the shape is *determined* is not so easily explained!

The events of budding as observed in the light microscope are summarized diagrammatically in Figure 1. Consider an unbudded cell (1A) in a growth-supporting medium. This cell first enlarges as an ellipsoid (1B) and then a bud emerges (1C). The bud continues to increase and finally a septum appears (1D). Separation follows and yields mother (1E) and daughter (1F) cells. The large mother cell can undergo another round via the shorter path indicated in Figure 1, (i.e., E → C → D) whereas the daughter cell must first enlarge before a bud appears (i.e., it engages in the longer path F → A → B → C → D). We can mention here, and develop later, that the unequal distribution of mass between mother and daughter at separation is dependent on growth rate. The difference is much less pronounced under maximal rates of growth.[1,2] Cells can bud many times as evidenced by the multiplicity of bud scars on mature cells. The separation of mother and daughter cell does not always occur before new buds are initiated. This gives rise to aggregates of cells, a condition more common than not in nature. This complication is for the most part side-stepped by molecular biologists by judicious selection of strains that do not undergo clumping. On the other hand, the phenomena of aggregation and flocculation are of considerable import in industrial fermentations because they facilitate the preparation of cell-free products.

Knowledge of the structural properties of the wall, which can explain the ability to maintain shape,[4,5] does not allow one to predict that the cell will grow as an ellipsoid (rather than as a rod or some other shape), when the bud will emerge, or how big the bud will be when the septum is made. Wall synthesis is guided by as yet unelucidated mechanisms to form a particular geometry and moreover the geometry of wall synthesis is changed at specific times. Since cell shape is an inherited property (and can be changed by mutation), it is clearly under genetic control. The control must eventually be expressed at the wall, perhaps by influencing the sites, orientations, or constituents of wall synthesis. Since synthesis is mediated by enzymes it seems reasonable that the control of shape determination is in part expressed by influencing the number, distribution, activity, and orientation of enzymes involved in wall synthesis. The control of cell shape is also expressed at sites other than the wall. Thus the timing of bud emergence and of septation relative to the rate of mass increase will determine the cell's dimensions. The temporal order depends on the execution of a "start" signal, DNA

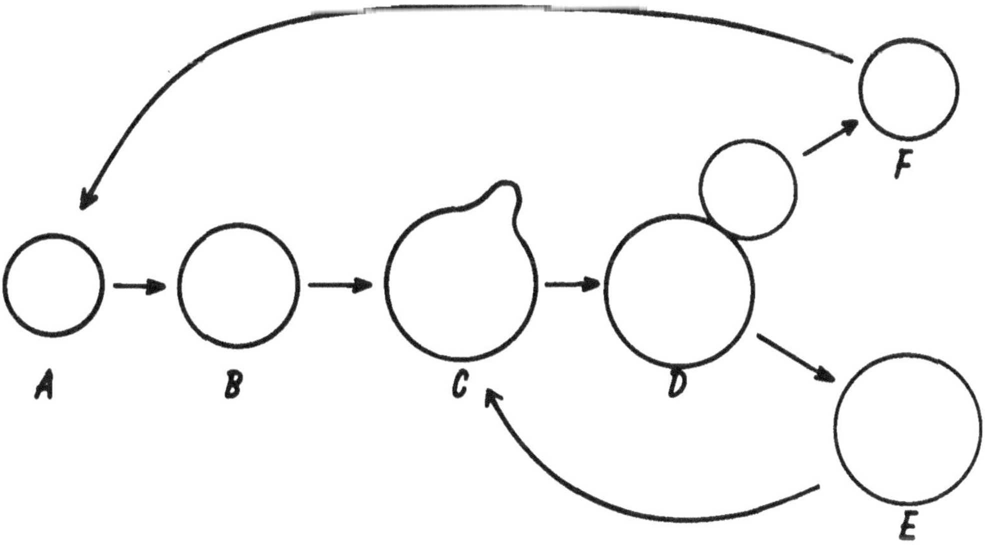

FIGURE 1. Diagrammatic representation of budding and unequal division in *Saccharomyces cerevisiae*.

synthesis, and other cell cycle events.[6] Thus, events occurring in the nucleus and cytoplasm are coordinated with, or somehow influence, those events at the wall that culminate in budding and division. A combination of ultrastructural, biochemical, and genetic approaches is called for in gaining an understanding of bud emergence and growth.

The success of the multidisciplinary approach is illustrated by studies of bacterial viruses which have resulted in a detailed understanding of their morphogenesis. Another beneficial aspect of this work has been the identification of intermediates of morphogenesis and the study of their processing.[7] For example, the phage head protein is assembled on a "core" substructure and the core is later removed.[8] The role of the core in morphogenesis could not have been learned from analyzing the mature head or by dissociating it and studying its reassembly in vitro. The distinction between shape determination and maintenance is particularly evident in phage morphogenesis. During tail assembly the sheath subunits are added sequentially in helical array around a preformed central tube of given diameter and length. At intermediate stages the process is reversible in the sense that the subunits are easily removed from the central core. However, when the last subunit is in place, the sheath is "fixed" so that the subunits are no longer easily removed even under harsh treatments.[9]

The detailed knowledge of the morphogenesis of bacteriophage P22 at the molecular level has, in fact, been used as a model to develop and verify procedures with yeast.[10,11] There are, however, fundamental differences between viral and yeast morphogeneses. The virus does not develop from a small virus into a large virus, rather the mature organism is assembled from its components. In this process it is possible to construct a substructure such as the head core or central tail tube, and to add in separate steps the shape maintaining outer components. On the other hand, yeast, which does develop from a "baby cell" into a "mature cell", must maintain a fully functional metabolism and structural integrity at all stages of its development. At any stage in the cell cycle, the internal contents of the cell can be extracted and the continuous wall maintains the identity of both mother cell and bud shapes.[5] Thus, the wall is able to maintain its shape at the same time wall synthesis is occurring and under the influence of those mechanisms that guide its geometry. There is no opportunity to construct a mature shape-determining substructure and then to assemble the rigid structural component as in the viral case. This coordination of determination and maintenance, in contradistinction to the separate stages of construction found in phage, explains the difficulties in identification of distinct intermediates and in the study of their processing in a cell such as yeast. This is not to exclude the possibility that the wall material produced at the sites of synthesis may be processed before the final network is arranged in a properly formed and complete wall.

The genetic approach has been fruitful in studies of molecular processing with phages[8-10] and this approach has also been useful in identifying the interrelationships of intermediate cell cycle events in yeast to which wall synthesis must be coordinated.[6] Analysis of yeast cell cycle mutants indicates that a nuclear entity may be involved in this coordination, especially with respect to nuclear events and bud emergence.[12] The genetic approach may also help to clarify the role of enzymes in the various aspects of yeast morphogenesis. When the enzymes and their effectors are fully described, the role of various reactions in the synthesis and perhaps processing of wall material may follow. A biochemical and ultrastructural comparison of the products formed in vitro by purified enzymes to those formed by protoplasts (see Volume II, Chapter 5) and those formed in vivo by yeast cells could help in understanding the role of enzymes in determining wall structure and shape, and perhaps aspects of morphogenesis that enzymes cannot explain. Mutations in the structural and regulatory genes for these enzymes would also help in identifying the role of enzymes in shape determination. It is not clear at this time whether the information for cell shape, and thus the identity of the bud, is contained only in genes functioning in envelope synthesis or whether some genes function in the synthesis of envelope material and others function in guiding the geometry of these processes. Notwithstanding some of these problems, yeast has a wall which influences its shape; it is amenable to biochemical and ultrastructural studies; and the species has a genetic system suitable for experimental manipulation. The emergence of a bud represents a change in the geometry of wall synthesis which is coordinated with other aspects of cellular development. Furthermore, yeast budding is a suitable model system amenable to the various approaches needed to gain insight into cellular morphogenesis generally. Some progress has been made in all of the major aspects outlined above.

II. BUD EMERGENCE

A. Genetic Control and Evolution

The interdependencies of cell cycle events have been defined by Hartwell and his collaborators[6] using temperature sensitive cell division cycle mutants (cdc mutants). The events originally used as cell cycle landmarks were bud emergence (BE), the initiation of DNA synthesis (iDS), DNA synthesis (DS), nuclear division (ND), cytokinesis (CK), and cell separation (CS). Bud emergence was found to depend on the functions of at least two genes, cdc 24 and cdc 28.[6,13] The cdc 24 mutants continue protein synthesis at the restrictive temperature (36°C) and undergo multiple rounds of DNA synthesis and nuclear division, but do not bud or divide.[6] Other mutants demonstrated that bud emergence, which in the wild type occurs at about the same time as the initiation of DNA synthesis, will occur even in the absence of DNA synthesis.[14] Thus, bud emergence does not depend on DNA synthesis and DNA synthesis does not depend on bud emergence. Mutations in cdc 28 prevent both bud emergence and DNA replication.[13] Therefore, budding and DNA synthesis, which are causally independent, are coordinated by the function of cdc 28 which is executed before either of the landmark events. Furthermore, cdc 4 mutants undergo bud emergence once every generation time in the absence of DS, ND, or CK and this indicates that the timing of budding can be dissociated from other cell cycle events.[14] The causal independence of bud emergence from DNA synthesis perhaps indicates that budding evolved and was integrated into the cell cycle after the evolution of the causally dependent pathway DS → ND → CK.[6] A sequencing procedure,[10] whose rationale was verified by studies with P22 phage, demonstrated that a "start signal" is identical to or codependent with the cdc 28 gene function that commits cells to a mitotic cell cycle.[11] These observations have been interpreted as indicating that the cdc 28 function did not evolve as a mechanism to integrate budding with DNA synthesis in yeast. Instead, the start function may have been part of a preexisting sequence:

start → DS → ND → CK

common to eukaryotic cells.[6] Subsequent work has revealed a "start" or commitment function in mammalian cells.[15]

The nature of sequences of functions leading to an observable marker event such as budding has been emphasized in the studies of cdc mutants. For example, the sequencing procedure referred to above demonstrated that the functions of three genes, cdc 4, cdc 7, and cdc 28 occur in the causally dependent sequence:

cdc 28 → cdc 4 → cdc 7

which leads to the initiation of DNA synthesis.[11] Also, "execution point" was defined as the time at which completion of an ongoing cell cycle becomes resistant to the restrictive temperature. This was interpreted to be the time that the function of the mutated gene is normally expressed, and when passed at the permissive temperature would allow the cell cycle to complete without further dependence on the function. For instance, the execution point of a ts DNA polymerase mutant would be the completion of DNA synthesis, after which the cell cycle would no longer depend on the function of that polymerase. The "diagnostic landmark" was defined as the first cell cycle landmark event affected by the mutation. The "terminal phenotype" was defined as the uniform morphology with which cells arrest. Definite "execution points" and "terminal phenotypes" were the original criteria of a cdc mutant.[6] The usefulness of these operational definitions can be illustrated using as a model cdc 7, the last gene whose function is specifically required for iDS. Sequential expression of the cdc 28 and cdc 4 genes leads to expression of the cdc 7 gene function. The execution time of cdc 7 is the initiation of DNA synthesis.[14] The terminal phenotype is a cell with a mature bud and undivided nucleus. In this example, the sequencing procedure defined the causally dependent pathway:

cdc 28 → cdc 4 → cdc 7

The execution point defined the time at which the cdc 7 function is required in the cell cycle. The terminal phenotype defined the interdependence of events, i.e., iDS is not required for BE, but is required for ND, CK, and CS.

These conclusions are based on measurements of the initial landmark which is the initiation of DNA replication, and illustrate the dependence of this type of analysis on known, measurable cell cycle characteristics. Suppose for the sake of argument that nuclear staining and procedures for measuring DNA synthesis were unknown when the study began. A series of mutants would have been found which would arrest at division with a mature bud. If nuclear staining then became available, some of these cell division mutants would have been termed nuclear division mutants. If measuring DNA synthesis then became possible, some of the nuclear division mutants would have been termed DNA synthesis mutants. The mutants termed nuclear division or DNA synthesis mutants, which were originally defined by measurements of macromolecular synthesis and morphology at the light microscope level, are under investigation at the ultrastructural and enzymatic levels so the names of the initial landmarks will probably change. In fact, enzymatic studies have demonstrated that mutants in cdc 9 are ligase mutants.[16] More relevant to this chapter are ultrastructural studies and genetic sequencing procedures demonstrating that the start signal required for bud emergence is codependent with duplication of a plaque associated with the nuclear membrane. Mutants such as cdc 28 were called plaque duplication (PD) mutants. Obviously, the decision to use a combined genetic, biochemical, and ultrastructural approach is not one which can be made or discarded at will. Regardless of changes of the terms used to describe the initial landmark, the mutants will remain cell division cycle (cdc) mutants which affect specifically single functions required for cell cycle progress. The genetic approach has defined interdependencies of cell cycle events, identified the existence of partially dependent and independent pathways, and concentrated attention on various aspects of cellular development. The commitment of division is receiving particularly intensive study.

B. Ultrastructure

As stated above, a nuclear plaque plays a role in bud emergence. The bud emerges at a site on the wall near the plaque and several observations demonstrate that duplication of this plaque is a prerequisite for bud emergence.

1. Mutants such as cdc 28 do not bud at the restrictive temperature (36°C), and the plaque does not duplicate.
2. When cdc 1 mutants arrest at 36°C, some of the cells have small buds and duplicated plaques; other cells have no buds and unduplicated plaques.
3. There is a prominent double plaque in cdc 4, which buds repeatedly in the absence of DNA synthesis or nuclear division.[12]

Microtubular spindles emanating from the plaque also appear to have a role in budding and perhaps its coordination with DNA synthesis. Microtubules extend from the plaque through the cytoplasm to the site of bud emergence. Spindles also connect the halves of the duplicated plaque which separate during DNA synthesis and nuclear division. The formation of a "complete spindle" in the nucleus, which does not occur in cdc 4, is required for the initiation of DNA synthesis. In fact, studies with cdc 7 indicate that the requirement of protein synthesis for the initiation of DNA synthesis is completed with the formation of a complete spindle. Thus, duplication of nuclear plaque is required for bud emergence and DNA synthesis. Also, spindles which extend from the plaque to the site of bud emergence may serve to coordinate this event with events in the nucleus associated with changes in other spindles emanating from the same plaque structure.[12]

Vesicles accumulate in the cytoplasm in the vicinity of the spindles at the site of the bud.[17] They appear to fuse with the plasma membrane and speculations about their function suggest that they may contain material for bud growth or enzymes which cause localized weakening of the wall. Hydrostatic pressure exerted on such a weakening was first suggested as the mechanism which initiates bud emergence,[18] and this would be consistent with the preferential appearance of buds at the sites of maximum wall curvature.[19] However, the thickness and ultrastructure of the wall does not change perceptibly during bud emergence[20] so that if a local change in the wall is associated with bud emergence, the change must be subtle and short lived, or undetectable by contemporary methodology.[21,22]

A ring of microfilaments appears at the site of bud emergence and remains at the base of the neck between the mother cell and bud until cytokinesis, at which time it disappears. Mutants defective in cytokinesis lack these microfilaments when they arrest. All other mutants with buds retain the microtubular ring.[23]

A ring of chitin appears at the neck between the mother cell and bud near the time of bud emergence.[24,25] However, there is no causal relationship between chitin synthesis and the emergence or growth of the bud since these events are not affected by inhibiting chitin synthesis with polyoxin D.[26] Also, inhibition of budding in cdc 24,[27] or by addition of a mating factor to wild type[28], does not prevent the synthesis of chitin, but does affect its distribution in the wall. Chitin synthesis is certainly required for septation. Cells are prevented from making functional septa when chitin synthesis is inhibited with polyoxin D.[26] This is consistent with evidence that the primary septum is chitin.[29] The dependence is not reciprocal, i.e., chitin synthesis is required for division but division is not required for chitin synthesis. Mutants arrested with an undivided nucleus in the neck between the mother cell and bud continue making chitin.[25]

That chitin synthesis, which begins near the time of bud emergence, has no apparent role in bud emergence but does function in septation is interesting, because the two processes are associated with the ability to divide. The emergence of a bud marks a change in wall synthesis from a geometry associated only with mass increase to a geometry associated with division, as will be discussed below.

C. Nutrition

The work discussed so far was done mainly with yeast populations growing at or near the maximum rate. In this case the bud grows to the size of the mother cell so that division results in two approximately equal cells. Cells growing at lower rates divide before the bud reaches the size of the mother.[1,2] After division a characteristic bud scar containing the chitin ring and primary septum remains with the mother. Also characteristic, but less prominent, is the complementary birth scar on the daughter cell. By accentuating these scars with fluorescence brighteners it is possible to evaluate the composition of such populations. At low growth rates (produced in a chemostat by limiting glucose) it was found that cells without bud scars varied more in both size and generation time than cells with bud scars. The maximum size and minimum generation time of the zero bud scar class were close to the uniform values for the mother cells. These results, combined with measurements of the size of cells at bud emergence, indicated that the small daughter cell required time to grow to the size of its mother cell before it budded, whereas the mother cell budded again soon after division.[1]

This is consistent with the observation that populations with long generation times have high frequencies of unbudded cells, and that the interval preceding bud emergence accounts mainly for the increased generation time.[30] Time lapse photography of populations with growth rates altered by various concentrations of cycloheximide was used in a recent study[2] to demonstrate unequal division and to follow the fates of resultant cells. In agreement with the above mentioned studies, it was found that as the generation time increased, the difference in size between mothers and daughters at division increased, and that the smaller daughter cells had to grow for a longer time before budding. Although the interval of time preceding bud emergence accounted for most of the variation in generation times, there was also some variation in the interval between budding and division. Even the mother cell with one bud scar grew somewhat before budding again.[2]

In all of the above studies unequal division was observed at low growth rates and at all growth rates an emerging bud grew to approximately the size of the mother cell before it in turn gave rise to a new bud. Under adequate nutrition most of this growth occurs while the bud is attached to the mother cell. But under nutritionally deprived conditions the bud separates from the mother before it "matures" and must continue growing as a single cell before it buds. The differences in generation time caused by nutritional conditions are consistently found to be accounted for mainly by the interval preceding bud emergence. Furthermore, when growth rate was varied by using media with different carbon sources, by limiting glucose concentration in a chemostat, or by using cycloheximide, the interval preceding bud emergence was found to correspond to the interval preceding DNA synthesis and the start signal, regardless of the generation time.[31,33] Thus cells without buds have not initiated DNA synthesis nor have they been committed to a cell cycle. It is the timing of this commitment which regulates the rate of budding and division in response to nutritional conditions.

In populations increasing in mass and cell number, it is not progress toward the start signal that limits the rate of mass increase. Instead, the rate of mass increase limits the progress toward the start signal and thus limits the frequency of bud emergence and division. Thus, it is the time required to reach a certain mass that determines the time a bud emerges.[34] Studies with inhibitors and mutants indicate that a cell measures its mass by counting the number of rRNA genes which are transcribed.[35] Collectively these results explain the common observations that (1) when yeast cultures approach stationary phase and the growth rate decreases, the frequency of unbudded cells increases and (2) the frequency of unbudded cells is high in stationary phase cultures—approaching 100% when the cultures remain in stationary phase for long intervals.[34,36] Clearly, as the growth rate decreases, the number of budded cells that divide becomes progressively larger than the number of previously unbudded cells that initiate a new bud.

More recent work demonstrated that suddenly removing any of a wide variety of essential nutrients from an actively proliferating culture allows budded cells to bring division to fruition while the emergence of new buds is prevented and net mass increase stops. Axiomatically, *an unbudded cell has never been observed to divide*. This is not trivial since cdc 24, which fails to bud at 36°C, continues to increase in size and undergoes multiple cycles of DNA synthesis and nuclear division.[6,27] In sum, a small unbudded cell must grow in order to put forth a new bud; a budded cell can divide without growing; and an unbudded cell cannot divide whether or not it is growing. Thus, the emergence of a bud marks a change in the wall from a spheroidal geometry accommodating the mass increase required for budding (but which cannot accommodate septation) to a state which can accommodate septation even if further mass increase is prevented.

III. BUD GROWTH

Measurement of cell size and wall synthesis consistently demonstrate that the bud grows and the mother does not.[37,38] However, it is only the envelope which is older in the mother cell than in the bud since a barrier restraining movement of newly made cytoplasmic material does not occur until late in the cell cycle.[39]

Most of the studies of bud growth have concentrated on the wall. The wall is made of glucan, mannan, chitin, and protein.[40] (See Volume I, Chapter 5 and Volume II, Chapter 2, for the chemistry of the cell wall.) Cell shape is maintained after extraction of the mannan and protein.[5] Chitin synthesis functions in septation but not in bud emergence or growth.[6] Therefore, it is the glucan that maintains the shape of the cell and thus the identity of the bud. Yeast glucan consists of two components; the more prevalent component is a β 1 → 3-linked glucose polymer with some β 1 → 6 linkages.[41] The other component has mostly β 1 → 6 linkages and some β 1 → 3 linkages.[47] The more prevalent component has been postulated as composing an inner fibrillar network that is covered by the more amorphous other component.[43] However, there is no direct evidence for this disposition in intact cell walls. Experiments with autoradiography and synchronous cultures demonstrated that wall synthesis occurs throughout most of the cell cycle, but a pause in glucan and mannan synthesis occurs near the time of division.[44] Since rapidly growing populations were used in which cells bud again soon after division, it was not clear whether the postulated pause occurs at the end of one cell cycle or the beginning of the succeeding one.

The topology of wall synthesis has been studied with a variety of methods, all of which demonstrate that synthesis occurs only in the bud and not the mother. (It is somewhat paradoxical, however, that those methods with high resolution have been applied more to the mannans than the glucans.) Pulse chase experiments indicate that there is no significant turnover, so material made in the bud is not later found in the mother cell (and vice versa).[45] When fluorescein-labeled concanavalin A was used to follow mannan synthesis, it was found to occur mainly at the tip of the bud, but also at its base.[46] Autoradiographic studies gave somewhat conflicting results which indicated mannan synthesis over the entire surface of small buds, mainly at the tip during the middle of the cell cycle and at the tip and sides of older buds.[47]

Glucan synthesis has been indicated to occur over a wide area of the bud[48] although culture in the presence of 2-deoxyglucose was found to cause lysis at the tip of the bud.[49] This last result also suggests that wall formation results from a balance between synthetic and lytic processes. According to this working hypothesis, inhibiting the synthetic reaction(s) allows the lytic process to weaken the wall and cause lysis. During normal wall synthesis, lytic enzymes act on the mechanical elements of the wall producing gaps which are filled by biosynthetic enzymes. The lytic and synthetic processes must be carefully coordinated since the wall must be able to maintain the shape of the cell at all stages of the cell cycle.[49] Protoplasts of yeast have been found to synthesize and secrete certain components of normal envelopes (see Volume II, Chapters 1 and 5) and to generate an insoluble fibrillar network[50] (see Volume II, Chapter 6). The formation of the wall does not at first reflect the shape of the cells from which the protoplasts were produced and that characteristic shape of the cell does not appear until several generations have occurred.[51]

The synthesis of glucan from UDP-glucose has been detected in permeabilized cells[52] and the in vitro synthesis of glucan has been recently described using particulate preparations.[53,54] In one case,[54] a highly active system from plasma membranes exhibited reversible activation by a nucleotide-like effector. Previous success in solubilizing chitin synthetase[55] indicates the possibility of future studies with solubilized glucan synthetase. Critical comparisons among soluble preparations, particulate preparations, permeabilized cells, resting or regenerating protoplasts, and intact cells will undoubtedly contribute to gaining an understanding of structural polysaccharide synthesis and of bud growth.

IV. SEPTUM FORMATION

The timing of division relative to the rate of mass increase in the bud determines the size of the daughter cell. The three aspects of morphogenesis involved in septation are *timing*, *placement*, and *orientation* of the septum. Studies by Cabib and his collaborators[29] demonstrate that the primary septum is chitin. Chitin synthesis was found to be required for septum formation, but not for bud emergence or growth.[26]

Chitin synthetase is located in the plasma membrane as an inactive zymogen[56] and this form of the enzyme has been solubilized,[55] thus eliminating the possibility of an artifactual activation by, for example, exposing a latent enzyme buried in the membrane. The zymogen is apparently distributed over the entire surface of the membrane, rather than in a localized position. Accordingly, localized activation of the zymogen has been postulated[57] as crucial to determining the position of chitin synthesis.

A vacuolar protease (proteinase B) was subsequently shown[58] to activate chitin synthetase. However, mutants able to form septa but lacking this specific activity have been isolated, thus indicating that proteinase B is not the activator of the zymogen in vivo.[59-61] In any event it is possible that the physiological activator is contained in the vacuole or some other compartment which is guided to the proper site on the membrane to activate the zymogen. The mechanisms that control *timing* of chitin synthesis are not known although it is found to be discontinuous in the cell cycle. The mechanism that *orients* the synthesis of chitin first as a ring around the collar between the mother cell and bud, and then in the transverse direction as a septum, is not understood. Either the sites of activation of the enzyme or the distribution of its product must be altered in the mutant cdc 24, which, as previously discussed, does not bud.[27]

V. CONCLUSIONS

When a new cell is produced by division in a slowly growing population it does not have a bud, it has not initiated DNA synthesis, it has not been committed to a cell cycle, and it cannot divide. The cell must grow to a critical size before a bud emerges. The rate of mass increase limits progress toward a start signal that is required for bud emergence, DNA synthesis, and commitment to a cell cycle. (This is different from progress toward the signal limiting the rate of mass increase.) The length of time needed to reach the mass required for bud emergence accounts for most of the effects of nutritional conditions on the rate of budding and division.

At or before execution of the start signal, information about the nutritional conditions in the environment and the metabolic state of the cell is integrated by various regulatory mechanisms including one which apparently counts the number of times rRNA genes are transcribed, which seems to be the cell's way of limiting the rate of mass increase. If the cell "decides" to become committed to a cell cycle based on this information, the start signal is executed. This signal is identical to, or codependent with, the duplication of a plaque associated with the plasma membrane. A bud subsequently emerges at a site near the nuclear plaque.

Spindles extend from the plaque to the site of bud emergence. These spindles, in conjunction with others which extend from the plaque into the nucleus, may coordinate events which initiate bud emergence at the wall and initiate DNA synthesis in the nucleus. Vesicles accumulate in the cytoplasm at the base of the bud as does a ring of microfilaments associated with the cytoplasmic membrane. A ring of chitin subsequently appears in the wall at the base of the bud. The possible appearance of these structures before bud emergence needs to be studied in slowly growing populations, in which cells must grow over an extended time interval before budding.

Should vital nutrients be removed, the cell can divide without further growth once the bud emerges. The cell cannot divide before bud emergence even if it undergoes multiple rounds of DNA synthesis and increases in mass under favorable nutritional conditions. Therefore, the appearance of a bud marks a change in the cell wall from a geometry that can accommodate only mass increase to a geometry that can accommodate mass increase and septation; or septation alone if mass increase is prevented by nutritional deprivation.

Once a bud emerges it is the only part of the cell that grows, in the sense that the mother cell does not exhibit net expansion. New wall formation, possibly involving balanced lytic and synthetic processes, occurs continuously throughout most of the cell cycle. Glucan is probably the most important structural component that maintains the shape of the cell but the organization of nascent glucan into its final form capable of shape maintenance is still far from understood. However, a series of experimental systems ranging in organization from intact cells, protoplasts, and permeabilized cells, to key particulate enzyme preparations, are now available. Soluble enzyme preparations and small molecule effectors may soon be at hand. Examining various degrees of organization of wall materials made by these systems should give insight into the processes which form functional walls from nascent wall polysaccharides. The mechanism by which the cell's shape is determined (and the identity of the bud is thus established) is entirely unknown. However these mechanisms almost certainly involve the distribution, orientation, and activation of membrane bound enzymes in those processes that organize nascent wall polysaccharide into the final functional form.

REFERENCES

1. **Beran, K.,** Analysis of growth and multiplication of yeast cells *Saccharomyces cerevisiae* in continuous culture from the aspect of relative age of cells, in *Proc. 4th Int. Symp. Yeasts*, Malek, I., Beran, K., Fencl, Z., Munk, V., Ricica, J., and Smrkova, H., Eds., Academic Press, New York, 1969, 87.
2. **Hartwell, L. H. and Unger, M. W.,** Unequal division in *Saccharomyces cerevisiae* and its implications for the control of cell division, *J. Cell Biol.*, 75, 422, 1977.
3. **Eddy, A. A. and Williamson, D. H.,** A method of isolating protoplasts from yeast, *Nature (London)* 179, 1957.
4. **Bacon, J. S. D., Farmer, V. C., Jones, D., and Taylor, I. F.,** The glucan components of the cell wall of bakers yeast *(Saccharomyces cerevisiae)* considered in relation to its ultrastructure, *Biochem. J.*, 114, 557, 1969.
5. **Northcote, D. H. and Horne, R. W.,** The chemical composition and structure of the yeast cell wall, *Biochem. J.*, 51, 232, 1952.
6. **Hartwell, L. H., Culotti, J., Pringle, J. R., and Reid, B. J.,** Genetic control of the cell division cycle in yeast, *Science*, 183, 46, 1974.
7. **Frankel, F. R.,** DNA replication after T4 infection, *Cold Spring Harbor Symp. Quant. Biol.*, 32, 485, 1968.
8. **Earnshaw, W. and King, J.,** Structure of phage P22 protein aggregates formed in the absence of the scaffolding protein, *J. Mol. Biol.*, 123, 721, 1978.
9. **Kikuchi, Y. and King, J.,** Assembly of the tail of bacteriophage T4, *J. Supramol. Struct.*, 3, 24, 1975.
10. **Jarvick, J. and Botstein, D.,** A genetic method for determining order of events in a biochemical pathway, *Proc. Natl. Acad. Sci., U.S.A.*, 70, 2046, 1973.
11. **Hereford, L. M. and Hartwell, L. H.,** Sequential gene function in the *Saccharomyces cerevisiae* DNA synthesis, *J. Mol. Biol.*, 84, 445, 1974.
12. **Byers, B. and Goetsch, L.,** Behavior of spindles and spindle plaques in the cell cycle and conjugation of *Saccharomyces cerevisiae, J. Bacteriol.*, 124, 511, 1975.
13. **Hartwell, L. H.,** Three additional genes required for DNA synthesis in *Saccharomyces cerevisiae, J. Bacteriol.*, 115, 966, 1973.
14. **Hartwell, L. H.,** Genetic control of the cell division cycle in yeast II genes controlling DNA replication and its initiation, *J. Mol. Biol.*, 59, 183, 1971.
15. **Pardee, A. B.,** A restriction point for control of normal animal cell proliferation. *Proc. Natl. Acad. Sci. U.S.A.*, 71, 1286, 1974.
16. **Johnston, L. H. and Nasmyth, K. A.,** *Saccharomyces cerevisiae* cell cycle mutant 9 is defective in DNA ligase, *Nature* (London), 274, 891, 1978.
17. **Moor, H.,** Endoplasmic reticulum as the initiator of bud formation in yeast *(S. cerevisiae), Arch. Mikrobiol.*, 57, 135, 1967.
18. **Nickerson, W. J.,** Molecular basis of form in yeasts, *Bacteriol. Rev.*, 27, 305, 1963.

19. **Freifielder, D.,** Bud position in *Saccharomyces cerevisiae, J. Bacteriol.*, 80, 567, 1960.
20. **McClary, D. O. and Bowers, W. D.,** The integrity of the cell wall during bud formation in yeasts, *Can J. Microbiol.*, 11, 447, 1965.
21. **Marchant, R. and Smith, D. G.,** Bud formation in *Saccharomyces cerevisiae* and a comparison with the mechanism of cell division in other yeasts, *J. Gen. Microbiol.*, 53, 163, 1968.
22. **Sentandreu, R. and Northcote, D. H.,** The formation of buds in yeast, *J. Gen. Microbiol.*, 55, 393, 1969.
23. **Byers, B. and Goetsch, L.,** A highly ordered ring of membrane associated filaments in budding yeast, *J. Cell. Biol.*, 69, 717, 1976.
24. **Hayashibe, M. and Katohda, K.,** Initiation of budding and chitin ring, *J. Gen. Appl. Microbiol.*, 19, 23, 1973.
25. **Cabib, E. and Bowers, B.,** Timing and function of chitin synthesis in yeast, *J. Bacteriol.*, 124, 1586, 1975.
26. **Bowers, B., Levin, G., and Cabib, E.,** Effect of polyoxin D on chitin synthesis and septum formation in *Saccharomyces cerevisiae, J. Bacteriol.*, 119, 564, 1974.
27. **Sloat, B. F. and Pringle, J. R.,** A mutant of yeast defective in cellular morphogenesis, *Science*, 200, 1171, 1978.
28. **Sheckman, R. and Brawley, V.,** Localized deposition of chitin on the yeast cell surface in response to mating pheromone, *Proc. Natl. Acad. Sci. U.S.A.*, 76, 645, 1979.
29. **Cabib, E. and Bowers, B.,** Chitin and yeast budding, localization of chitin in yeast bud scars, *J. Biol. Chem.*, 246, 152, 1971.
30. **von Meyenburg, H. K.,** The budding cycle of *Saccharomyces cerevisiae, Pathol. Microbiol.*, 31, 117, 1968.
31. **Bradford, J. P. and Hall, B. J.,** Estimation of the length of cell cycle phases from asynchronous cultures of *Saccharomyces cerevisiae, Exp. Cell Res.*, 102, 276, 1976.
32. **Slater, M. L., Sharrow, S. O., and Gart, J. J.,** Cell cycle of *Saccharomyce cerevisiae* growing at different rates, *Proc. Natl. Acad. Sci., U.S.A.*, 74, 3850, 1977.
33. **Carter, B. L. A. and Jagadish, M. N.,** Control of cell division in the yeast *Saccharomyces cerevisiae* cultured at different growth rates, *Exp. Cell Res.*, 112, 373, 1978.
34. **Johnston, G. C., Pringle, J. R., and Hartwell, L. H.,** Coordination of growth with cell division in the yeast *Saccharomyces cerevisiae, Exp. Cell Res.*, 105, 79, 1977.
35. **Singer, R. A., Johnston, G. C., and Bedford, D.,** Methionine analogs and cell division in the yeast *Saccharomyces cerevisiae, Proc. Natl. Acad. Sci., U. S.A.*, 75, 6083, 1978.
36. **Johnston, G. C.,** Cell size and budding during starvation of the yeast *Saccharomyces cerevisiae, J. Bacteriol.*, 132, 738, 1977.
37. **Mitchison, J. M.,** The growth of single cells. II. *Saccharomyces cerevisiae, Exp. Cell Res.*, 15, 214, 1958.
38. **Johnson, B. F.,** Morphometric analysis of yeast cells, *Exp. Cell Res.*, 39, 577, 1965.
39. **Hartwell, L. H.,** Genetic analysis of the cell division cycle in yeast. IV. Genes controlling bud emergence and cytokinesis, *Exp. Cell Res.*, 69, 265, 1971.
40. **Cabib, E.,** Molecular aspects of yeast morphogenesis, *Ann. Rev. Microbiol.*, 29, 191, 1975.
41. **Manners, D. J., Masson, J. C., and Patterson, J. C.,** The structure of a β-(1-3)-D-glucan from yeast walls, *Biochem J.*, 135, 19, 1973.
42. **Manners, D. J., Masson, J. C., Patterson, J. C., Bjorndal, H., and Lindberg, B.,** The structure of a β-(1,6)-D-glucan from yeast cell walls, *Biochem. J.*, 135, 31, 1973.
43. **Kopecka, M., Phaff, H. J., and Fleet, G. H.,** Demonstration of a fibrillar component in the cell wall of the yeast *Saccharomyces cerevisiae, J. Cell Biol.*, 62, 66, 1974.
44. **Beily, P.,** Changes in rate of synthesis of wall polysaccharides during the yeast cell cycle, *Arch. Microbiol.*, 119, 213, 1978.
45. **Kratky, Z., Beily, P. and Bauer, S.,** The wall mannan of *Saccharomyces cerevisiae*. Metabolic stability and release into growth medium, *Biochim. Biophys. Acta*, 404, 1, 1975.
46. **Tkaz, J. S. and Lampen, J. O.,** Wall replication in *Saccharomyces* species: use of fluorescein-conjugated concanavalin A to reveal the site of mannan insertion, *J. Gen. Microbiol.*, 72, 243, 1972.
47. **Farkas, V., Kovarik, J., Kosinova, A., and Bauer, S.,** Autoradiographic study of mannan incorporation into the growing cell walls of *Saccharomyces cerevisiae, J. Bacteriol.*, 117, 265, 1974.
48. **Beily, P., Kovarik, J., and Bauer, S.,** Cell wall formation in yeast, *Arch. Mikrobiol.*, 94, 365, 1973.
49. **Johnson, B. F.,** Lysis of yeast cell walls induced by 2-deoxyglucose at sites of glucan synthesis, *J. Bacteriol.*, 95, 1169, 1968.
50. **Necas, O.,** Cell wall synthesis in yeast protoplasts, *Bacteriol. Rev.*, 35, 149, 1971.
51. **Svoboda, A. and Necas, O.,** Morphogenesis during protoplast reversion in dimorphic yeast, in *Proc. 4th Int. Symp. Yeasts*, Malek, I., Beran, K., Fencl, A., Munk, V., Ricica, J., and Smrkova, H., Eds., Academic Press, New York, 1974, 215.

52. **Sentandreu, R., Elorza, M. V., and Villanueva, J. R.,** Synthesis of yeast wall glucan, *J. Gen. Microbiol.*, 90, 13, 1975.
53. **Balint, S., Farkas, V., and Bauer, S.,** Biosynthesis of β-glucans catalyzed by a particulate enzyme preparation from yeast, *FEBS Lett.*, 64, 44, 1976.
54. **Shematek, E., Braatz, J., and Cabib, E.,** Synthesis of yeast cell wall glucan and its regulation, *Fed. Proc. Fed. Am. Soc. Exp. Biol.*, 37, 1394, 1978.
55. **Duran, A. and Cabib, E.,** Solubilization and partial purification of yeast chitin synthetase, *J. Biol. Chem.*, 253, 4419, 1978.
56. **Duran, A., Bowers, B., and Cabib, E.,** Chitin synthetase zymogen is attached to the yeast plasma membrane, *Proc. Natl. Acad. Sci. U.S.A.*, 72, 3952, 1975.
57. **Duran, A., Cabib, E., and Bowers, B.,** Chitin synthetase distribution on the yeast plasma membrane, *Science*, 203, 363, 1979.
58. **Cabib, E. and Ulane, R.,** Chitin synthetase activating factor from yeast, a protease, *Biochem. Biophys. Res. Commun.*, 50, 186, 1973.
59. **Jones, E.,** Proteinase mutants of *Saccharomyces cerevisiae, Genetics*, 85, 23, 1977.
60. **Wolf, D. H. and Ehmann, C.,** Isolation of yeast mutants lacking proteinase B activity, *FEBS Lett.*, 92, 121, 1978.
61. **Zubenko, G. S., Mitchell, A. P., and Jones, E. W.,** Septum formation, cell division, and sporulation in mutants of yeast deficient in proteinase B, *Proc. Natl. Acad. Sci. U.S.A.*, 76, 2395, 1979.

Chapter 4

FISSION

E. Streiblová

TABLE OF CONTENTS

I. INTRODUCTION

Fission is a vegetative form of reproduction characterized by the division of a parental cell into approximately equal daughter cells. It is characterized by the development of a cross wall without cell constriction. Separation of daughter cells results from cleavage along the cross wall.

This chapter will deal exclusively with *Schizosaccharomyces pombe,* the best known unicellular eukaryote multiplying by binary fission. The essential features of its envelope topography will be summarized first. In the succeeding section, the essential questions of polar growth together with septum formation will be considered briefly. Recurring emphasis will be placed on the influence of cell cycle controls on cell wall extension and septation. These controls assure that both events occur in the correct sequence and at the right time during the normal cell cycle of *Schiz. pombe.*

II. ENVELOPE TOPOGRAPHY

Fluorescence microscopy as well as different methods of electron microscopy have been used to monitor cell wall structures of *Schiz. pombe* cells during growth and division. This approach has contributed to the understanding of structural changes of the cell envelope in almost all stages of fission.[1-6]

The most prominent features of the cell surface are division scars.[7] A micrograph of a freeze-fractured and etched cell is shown in Figure 1. Scars are formed at the sites of septum cleavage, the scar margin being formed from the lateral cell wall while the scar plug originates from the septal wall. Scar margins mark boundaries between wall areas of different cycles. Scar plugs are the sites of growth of the new cell wall. After a large number of divisions, cells are encountered with different numbers of division scars and these persist and segregate in the following generations of daughter cells (Figure 2).

The conserved central region of the cell envelope possesses cell wall rings of different age. As shown schematically in Figure 3, the oldest original wall ring is bounded on both sides by scar margins. In cells having more than two scars additional wall rings are formed. As distinguished from the original wall ring, the additional wall rings always have a plug wall band. Depending on the number of divisions, there are further additional wall rings added outside the existing ones. The surface enlargement during the cell cycle is due to the growth of the polar wall at one or both ends. Different regions of the envelope have different affinities for primulin (a fluorescent dye) as shown for a three-scarred cell in Figure 4.

III. CELL WALL GROWTH

The cell wall growth of *Schiz. pombe* may be analyzed with reference to the conserved central area of the cell surface which is inserted between the two poles. This cylindrical portion of the cell wall is nonextensile[7] and essentially does not increase after cell birth. In addition, each nascent cell contains a primary pole (P) conserved from the earlier division and a secondary pole (S) formed in the current cell cycle.[7,8] As will be considered later, the location of the septum is not primarily linked to the position of the conserved central area of the cell surface.

A. Wall Extension

Cells of *Schiz. pombe* double in length during the cell cycle. The extension rate increases for about three quarters of the cell cycle and then stops.[7-9] Microscopic[7] and immunofluorescent[10] observations, autoradiography[8], and time-lapse cinematography[9]

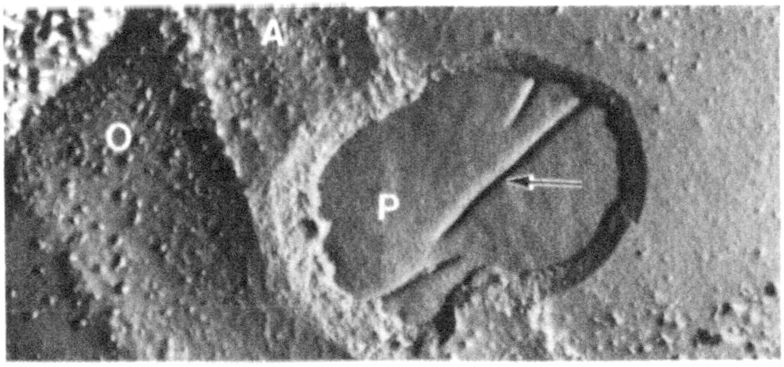

FIGURE 1. Electron micrograph of a freeze-fractured and etched cell of *Schizosaccharomyces pombe*. This cell displays three division scars. The original wall ring (O) and additional wall ring (A) are accentuated by deep-etching. Part of the cell wall has been fractured away and the protoplasmic half of the underlying plasma membrane (P) displays invaginations, one of which is marked with an arrow. Bar equals 0.5 μm.

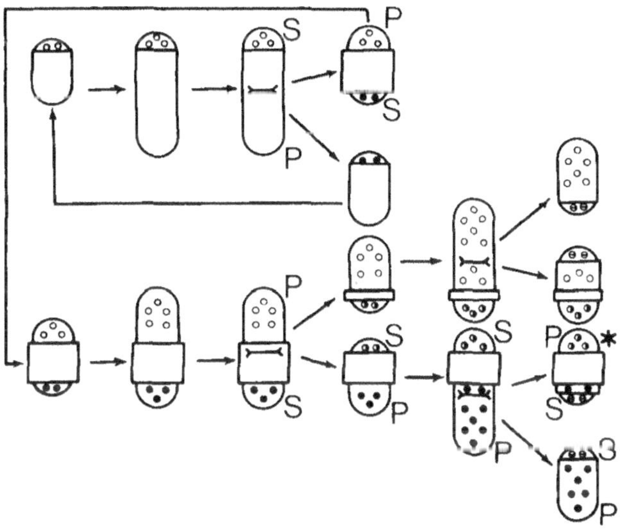

FIGURE 2. Schematic of fission process in *Schizosaccharomyces pombe*. A single-scarred cell (upper left) doubles its length by asymmetric extension at primary (P) and secondary (S) poles. After termination of growth a septum is formed at the equator. Cleavage at the septum gives rise to a second-generation cell with two scars and another with one scar. As continued in lower left, subsequent division produces two third-generation cells, each with two scars. The fourth-generation contains two three-scarred cells and two single-scarred cells. The oldest part of the cell surface is shown without internal symbols. The order of decreasing surface age is indicated by the following symbols: ○, ●, ◑, and ⊖. The scheme does not indicate relative proportions in a growing culture.

have been used to detect the area of new growth. These studies have provided convincing evidence for the highly localized growth of the cell wall in polar zones.

Contradictory results have been reported concerning the number of growing poles. Mitchison[7] originally attributed bipolar growth to an undefined fraction of the cell population. Johnson,[8] using autoradiography, concluded that the new wall is inserted predominantly at one pole in most cells. In contradistinction, our findings suggest a

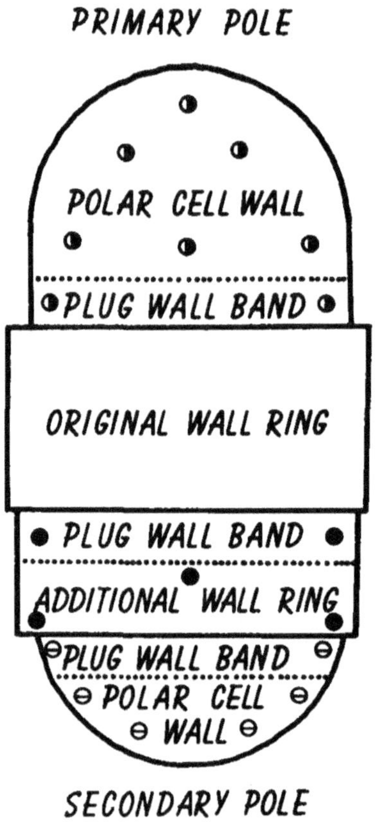

FIGURE 3. Diagram of a three-scarred cell equivalent to that indicated by (*) in Figure 2. Surface age is indicated by symbols as in Figure 2.

substantially lower proportion of cells growing only at a single pole. Our observations with cinematography indicate that in all cells growth at the older primary pole is responsible for cell extension, but in the great majority of cells there is also limited growth at the secondary pole. New cell wall generation at the secondary pole may be small but can be distinguished using fluorescence microscopy (Figure 5). Growth at the primary pole lasts throughout the entire stage of extension growth whereas the secondary pole elongates towards the end of this stage.[9]

Conflicting results[8] could be explained by differences in strains and growth conditions. However, a subsequent paper[11] from Johnson's group clearly shows that the previous discrepancy lay in the interpretation of fluorescence microphotographs. In the published photographs[11] a small fluorescent cap is visible at the secondary poles of cells containing septa. These observations had been interpreted in terms of unipolar growth probably because the caps were thought to be the result of rounding-out of the poles during the current division.

At the secondary poles division scars always constitute reliable references for measuring polar growth.[9] Unfortunately, division scars are not of equal importance for measuring the growth at primary poles because these regions often lack division scars. Poles growing from division scars always possess a boundary of minimum fluorescence between the growing tip and the nonextensile region. This structure, the plug band, is of considerable use for the understanding of the surface topography.

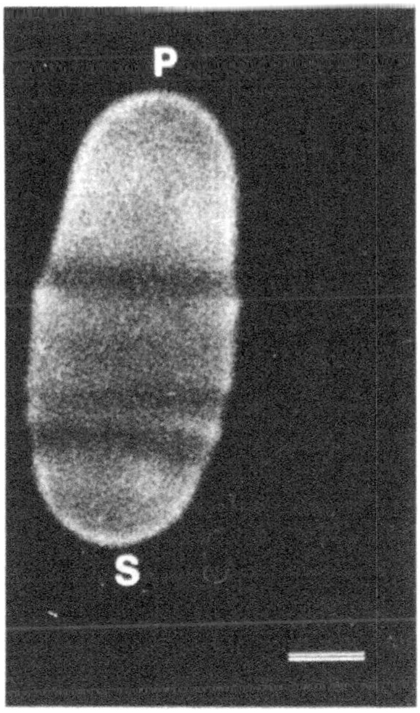

FIGURE 4. Fluorescence micrograph of a three-scarred cell of *Schizosaccharomyces pombe*. Wall rings and polar walls can easily be distinguished due to the minimum fluorescence of the plug wall bands. Primary (P) and secondary (S) poles are indicated. Bar equals 2 μm.

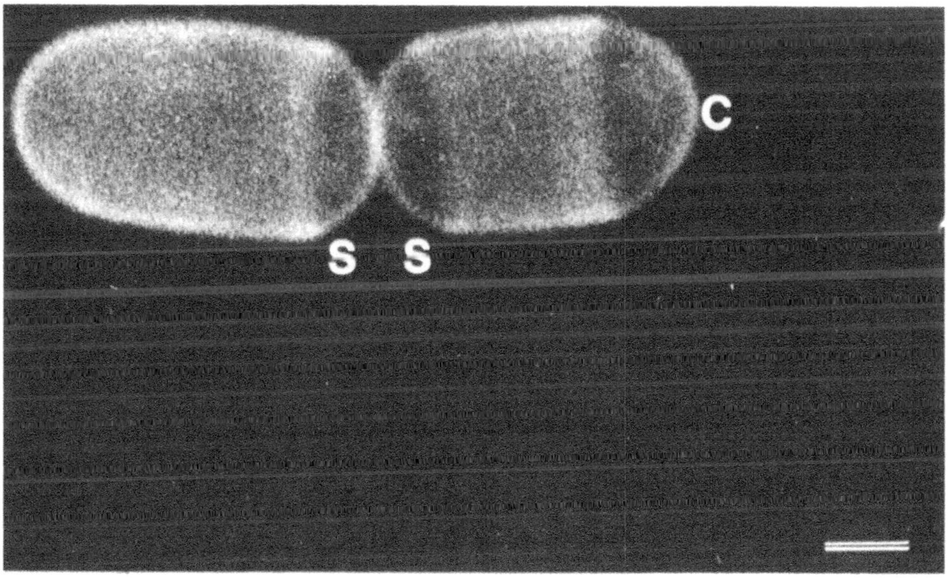

FIGURE 5. Fluorescence micrograph of two daughter cells of *Schizosaccharomyces pombe* immediately after septum cleavage. The secondary (S) poles show scar plugs with minimum fluorescence, i.e., at that time without growth. On the right-hand cell a small fluorescent cap (C) is visible indicating limited growth at this pole in the previous cell cycle. Bar equals 2 μm.

Although the pattern of polar extension can be clearly recognized in different experimental situations, its detailed relationship to the cell cycle remains obscure. The coordination of growth at the primary and secondary pole is subject to a controlling mechanism which is not well understood at present despite its considerable theoretical interest. The control appears to be size rather than time related.[12] A genetic approach to this problem appears in Section IV.

B. Septum Formation

To follow septum formation, use can be made of the high fluorescence of this structure when stained with primulin or Calcofluor®,[4,13] (available from Fluka, Switzerland and American Cyanamid, U.S., respectively). At the beginning of the constant volume stage[7] a bright ring appears at the perimeter of the cell equator. The ring corresponds to the initial stage of the septum wall. It gradually grows centripetally in the manner of a closing iris diaphragm. After the septum is completed, cell separation is accomplished by septum cleavage to yield daughter cells of equal volume. At the electron microscope level, septum formation is first discerned as an inward furrowing of the plasma membrane followed by deposition of septal wall material. The fully developed septum appears in thin sections as a symmetrical structure, consisting of two lamellae separated by an area of different electron opacity.[13-15]

Before proceeding, one can summarize the differences between the lateral and septal walls as follows:

1. The pattern of primulin staining indicates that the septum wall differs from the peripheral wall in microfibril assembly and orientation.[5]
2. Differential incorporation of 2-deoxyglucose and sensitivity to hydrolysis implicate differences in the biochemistry of septum biosynthesis and wall extension.[15]
3. The fact that mutants may be blocked specifically in septum formation[16] suggests an independent control of lateral wall vs. septal wall genesis.

IV. GENETIC CONTROL OF FISSION

A number of temperature-sensitive cell division cycle mutants (cdc mutants) of *Schiz. pombe* have been characterized.[16-18] While apparently normal when grown at 25°C these mutants become blocked at various stages of the cell cycle when grown at 35°C. Under these circumstances cell elongation and septum formation are no longer coordinated. Some of these mutants have defects in DNA replication, nuclear division, and early septum formation[16] and produce filamentous forms without substantial changes in cell diameter, i.e., the cell continues to elongate but septum formation is prevented.

Other mutants with defects in late events of septum formation,[16] and one mutant forming multiple septa,[18] are not affected with respect to cell elongation but form dumbbell-like forms with swollen poles, i.e., they accumulate high amounts of delocalized septal material.

A. Septum Induction

In *Schiz. pombe*, the positioning of the septum site is of fundamental importance to the cell. Although the process remains to be elucidated it must be subject to the following restraints. Let us first consider the equatorial location of the septum in relation to the cell surface. It is obvious that septum location does not originate in the preexisting cell wall topography. The old cell wall is not the primary target of the septation stimulus, as will be seen below, and the septum position does not exhibit any regular geometrical relationship to the nonextensile region. There may be some

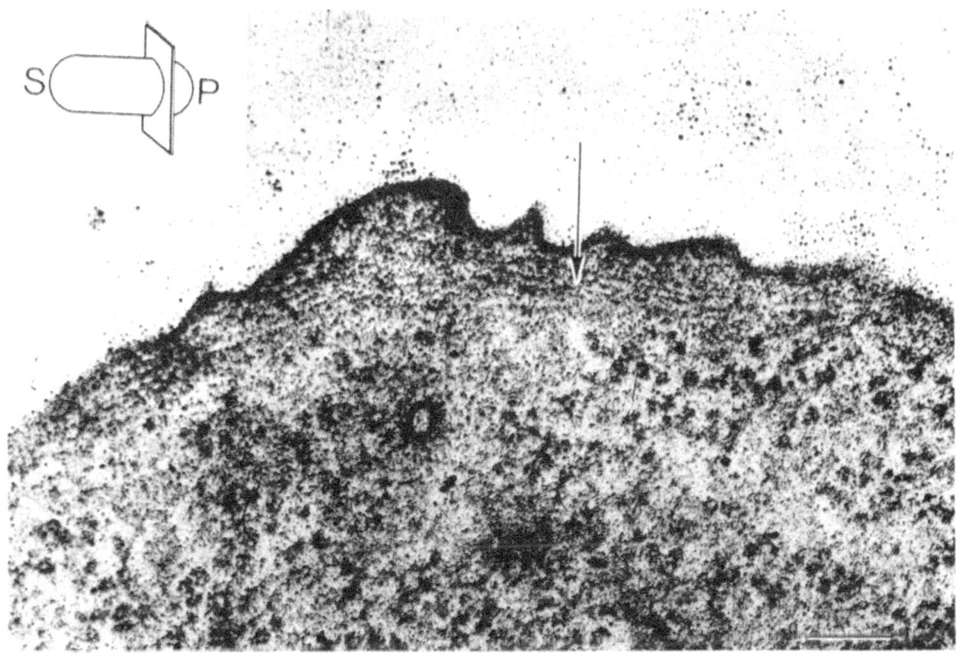

FIGURE 6. Ultrathin section (see insert) through the primary pole of the mutant cdc 12-112 of *Schizosaccharomyces pombe.* A bundle of microfilaments (arrow) approximately 6 to 7 nm in diameter is located at the perimeter of the primary pole. This bundle is associated with the plasma membrane and is oriented perpendicularly to the long axis of the cell. Bar equals 0.1 μm.

significance in the fact that the cells are not regular in shape and that the diameter of the cell wall cylinder increases at sites of old cell wall rings.[19] Therefore, a division of the cell into daughters of equal length will not result in daughters of equal volume. The analysis by Johnson et al.[11] indicates that the septum is so located as to ensure daughters of equal volume rather than equal length. Any model of septation control in *Schiz. pombe* must account for this phenomenon.

Observations with animal cells suggest that the positioning of the cytokinetic furrow is related to the position of the dividing nucleus.[20] Analysis of the behavior of some cell cycle mutants in *Schiz. pombe* also supports this suggestion. The timing of septation is closely related to mitosis[16] and the positioning of the septum is likely to be determined by the position of the dividing nucleus.[21] A circumferentially aligned bundle of microfilaments encircles the primary pole of the *Schiz. pombe* mutant cdc 12-112,[22] (Figure 6). At first sight, this bundle resembles the contractile ring which is induced by the mitotic nucleus and constricts the cleavage furrow in dividing animal cells.[23] The presence of a similar intracytoplasmic filamentous belt in fungi and its association with the plasma membrane of the developing septum was first pointed out by Girbardt.[24,25]

The factors important for septum induction[26] may be summarized as follows:

1. The cell size (mass or volume) governs the periodicity of nuclear division.
2. The position of the septum is determined during the oscillation period of nuclear division.
3. Septum formation is triggered only during meta-anaphase.
4. The target of the regulatory signal is the plasma membrane.
5. Once induced, septum formation continues independently of further displacement of nuclei.

6. In the period from anaphase to the termination of nuclear division, biochemical changes take place in the plasma membrane at the site of induction.
7. The first morphological sign of septum triggering only appears after termination of the nuclear division; at the site of induction a ring of microfilaments is assembled prior to any sign of furrowing of the plasma membrane; the position of the ring determines the position of the septum.

The above series includes factors common to both fungal and animal cytokinesis. In fungi, attention has so far been concentrated on *Trametes versicolor*,[24-28] a basidiomycete. Cytoplasmic microfilaments associated with septum development have now been found in other fungi.[29] In animal cells, the position of the microfilament ring in the cleavage furrow is probably the best documented example of an interaction between cell structures at the supramolecular level.[30]

The above data explain why cdc mutants of *Schiz. pombe* that block early mitosis also block cytokinesis and, as one shall see later, why cytokinesis takes place in the absence of nuclear division in cdc 16-116.

1. Late Septation Mutants

The terminal phenotypes of mutants in cdc genes 3,4,8, or 12 accumulate considerable amounts of septal material[16,31] whose location is contrasted with that of the normal septum in the wild type (Figure 7A). In the mutants, the primulin-staining material occurs in patches (Figures 7B and 7C) and is distributed irregularly within the cell envelope but outside the plasma membrane. The aberrant deposits occur in plasma membrane infoldings albeit with varying degrees of packaging. The cytoplasm of the mutants displays signs of altered secretory activity, including fenestration of endoplasmic reticulum and dictyosomes, dilation of secretory vacuoles, and accumulation of cytoplasmic vesicles. Mutants cdc 3-6 and cdc 12-112 were found to contain cytoplasmic microtubules extending singly or in groups through the aberrant septa from one cell pole to the other. In the later mutant, the author found that the primary pole was encircled by a bundle of cytoplasmic microfilaments associated with the plasma membrane and oriented at right angles to the microtubular axis.

Late septation mutants are interpreted as being defective in septum formation and, at the same time, as blocked in the constant length stage.[16] The cytological picture of the terminal phenotypes of these mutants is complex and the primary lesions of the mutants cannot as yet be determined. However, the mutations affect directly or indirectly the internal cytoskeleton[22] which is known in other cell types to participate in the control of cytokinesis as well as in the maintenance of cell shape.

The delocalized septum formation in the mutant cdc 12-112 may be due to a primary defect in the assembly and/or function of the ring of cytoplasmic microfilaments. The mutation could affect some of the steps of septum induction culminating in the formation of the ring of microfilaments. Alternatively, the filamentous ring could be obliterated in the first septum formation with only a second filamentous ring present. The mutant cells in which the ring was found, were blocked for 6 hr at the restrictive temperature. In these cells a second nuclear division (Figure 7D) must have taken place.[32]

2. Multiseptate Mutants

The mutant phenotype of the cdc 16 gene possesses some remarkable features. Cells shifted into restrictive conditions in the first three-fourths of the cell cycle form, after nuclear division, successively up to five septa between separated nuclei. Cells shifted in the last one-fourth of the cell cycle divide normally. One of the daughter cells then

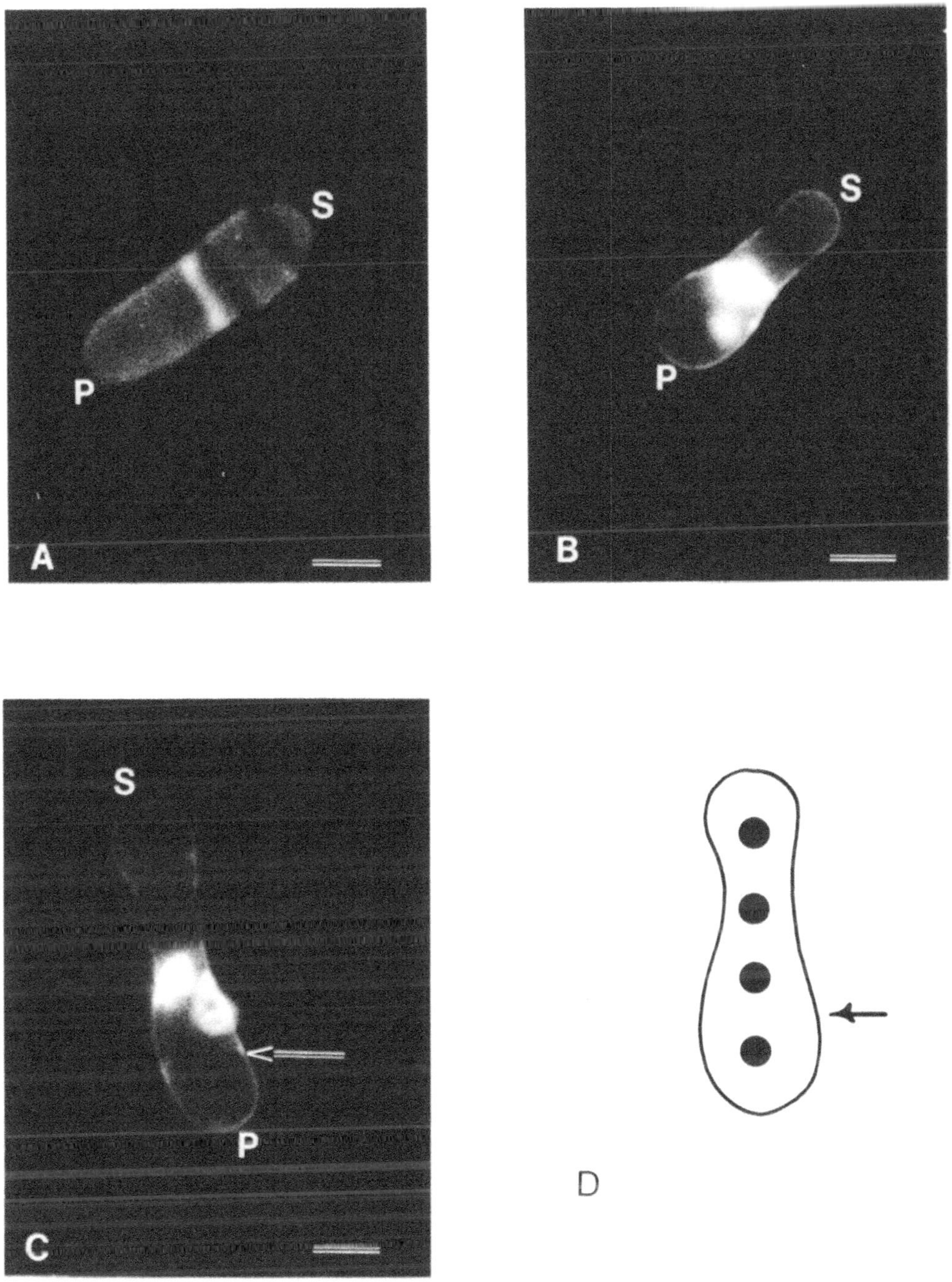

FIGURE 7. (A) Wild type cell of *Schizosaccharomyces pombe*. The septum is the only structure marked by bright fluorescence. (Primulin stained.) Bar equals 3 μm. (B) Mutant cdc 12-112 blocked 3 hr at 35°C. The cell shows bipolar growth with identifiable primary (P) and secondary (S) poles. It accumulates large quantities of highly fluorescent septal material in the equatorial region. Bar equals 3 μm. (C) Mutant cdc 12-112 blocked 6 hr at 35°C. Bright patches of the first delocalized septum are evident. After prolonged cultivation at this temperature the cells showed a second delocalized septum and, at the same time, the site (arrow) where the bundle of microfilaments encircling the primary pole was found in ultrathin sections. Bar equals 3 μm. (D) Diagram of the mutant cell depicted in (C). Under prolonged cell cycle arrest the cell completes the second nuclear division. The four nuclei (filled circles) from the second division are disposed as indicated. Probable location (arrow) of the microfilament bundle is indicated.

forms the above multiseptate phenotype while the other forms an asymmetric septum without any further nuclear division. Growth of the smaller anucleate compartment ceases. The nucleate cell develops into the multiseptate phenotype. Double mutants showed that the expression of this mutation is dependent on nuclear division.[18]

This interesting mutant defies the widely accepted notion that septum formation depends on the completion of nuclear division. A theoretical explanation of the behavior of this mutant was offered by Minet et al.[18] who feel that multiseptate cells are explained by two antagonistic controls that open and close the "competence window for septation." The function of the wild type cdc 16 gene product is required for the production of an inhibitor of further septation which closes this competence window. The formation of anucleate cell compartments is attributed to a partial thermal inactivation of the gene product cdc 16-116 immediately after the shift.

An alternative explanation is also possible. The cdc 16-116 mutant is altered in the induction of septation (Figure 8). The production of multiseptate cells may be interpreted in terms of an induction of septation in an early stage of nuclear division. In wild-type cells normally only a single septum is induced during the metaphase. In the cdc 16-116 mutant this normal control is disturbed and a number of septa may be induced in slightly different positions in the equatorial region. Successive inductions may be related to displacements of metaphase nuclei to excentric positions. In other fungal and animal examples[20,27,28,33,34] there is convincing evidence for multiple furrowing. The number of furrows and septa may exceed the number of mitotic apparatus.

The production of anucleate cell compartments in a predetermined position, topologically related to the septum from the earlier division, suggests that the plasma membrane was stimulated to induce septation in the absence of nuclear division. This suggestion apparently conflicts with other evidence which indicates that furrow and septum induction always requires dividing nuclei working in concert.[20,33] This conflict may be resolved by using the analogy with animals in which fragments of cells containing the presumptive furrow or otherwise intact cells deprived of their mitotic apparatus can divide.[20,34-36] Cells of cdc 16-116 shifted to restrictive conditions in the last quarter of the cell cycle show a normal first septum induction. The second septum found in the daughter cell is triggered by a secondary position of the mitotic apparatus. The first septum divides the cell before the second septum formation can take place. Further analogies with the situation in animal cells[30,33,35,36] are also possible.

B. Control of Surface Extension

Mutants with a primary defect in enzyme processes responsible for surface extension have not yet been described in the literature. The most likely candidates for genetic analysis of the growing cell wall would be enzymes concerned with turnover of β-glucans which represent the main morphogenic components of the *Schiz. pombe* wall.[37,38] Conditional mutants with defective glucan synthetase would eventually lyse under restrictive conditions since the putative hydrolytic enzymes in the cell wall would still be active (see Volume II, Chapter 1). Similar effects have been observed with 2-deoxyglucose.[39-42] On the other hand lysis would not occur in mutants deficient in glucan hydrolases. Such mutants would undoubtedly be very valuable.

Those cdc mutants of Nurse et al.[16] so far examined by primulin fluorescence have shown no obvious defects in the structure of the cell wall; hence orchestration of synthesis and breakdown of cell wall components is not seriously compromised during prolonged cell wall extension. Cell wall growth is coupled with the synthesis of macromolecules and may be related to the growth cycle[43] rather than to the DNA-division cycle. Filamentous cells of cdc mutants have identifiable nonextensile regions[16] but the pattern of polar growth remains to be elucidated. On the other hand,

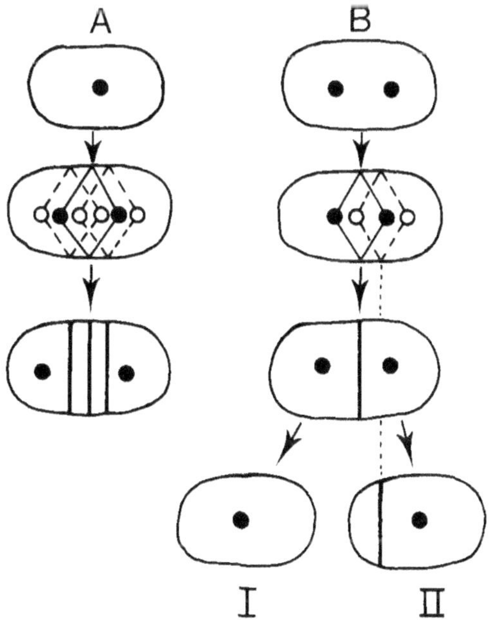

FIGURE 8. Diagrammatic representation of the behavior of mutant cdc 16-116 based on the proposal that septum position is determined in metaphase and that displacement of dividing nuclei induces displaced septa. (A) Cells shifted to restrictive temperature until metaphase exhibit a homogeneous phenotypic response and form three (sometimes as many as five) septa in the equatorial region. (B) Cells shifted to restrictive temperature at a later stage in the cell cycle undergo cell separation giving two dissimilar daughter cells. One daughter (I) develops into the multiseptate phenotype as in (A) whereas, the other daughter (II) has an asymmetric septum. Only the nucleate compartment of (II) develops into the multiseptate phenotype. The model also shows the primary (full lines) and displaced (dashed lines) positions of the mitotic apparatus in the mutant. As shown in sequence (B), septum induction may be subsequently fulfilled in the following cell cycle of the daughter thus producing an anucleate compartment as well as a nucleate one (II).

cytokinetic mutants which are not primarily defective in cell elongation[16,18] retain identifiable primary and secondary poles and the growth pattern remains unaltered.

In the phenotype of cdc 3-6 and cdc 12-112, cytoplasmic microtubules (Figure 9) are repeatedly found running throughout disorganized septa and terminating near the cell poles. It is quite possible that this arrangement of microtubules imitates the central strand of cytoplasm observed in Giemsa-stained preparations of the mutant cells.[16] In intact cells microtubules have been found only in regenerating protoplasts.[44] Although no proof is yet available it is the author's belief that the control of cell length is a microtubule-based phenomenon.

Mutant strains wee 1-1, wee 1-50, wee 2-1, cdc 2-M35, and cdc 25-22 vary in cell length at division over a three-fold range.[45,46] The initiation of growth at the two poles

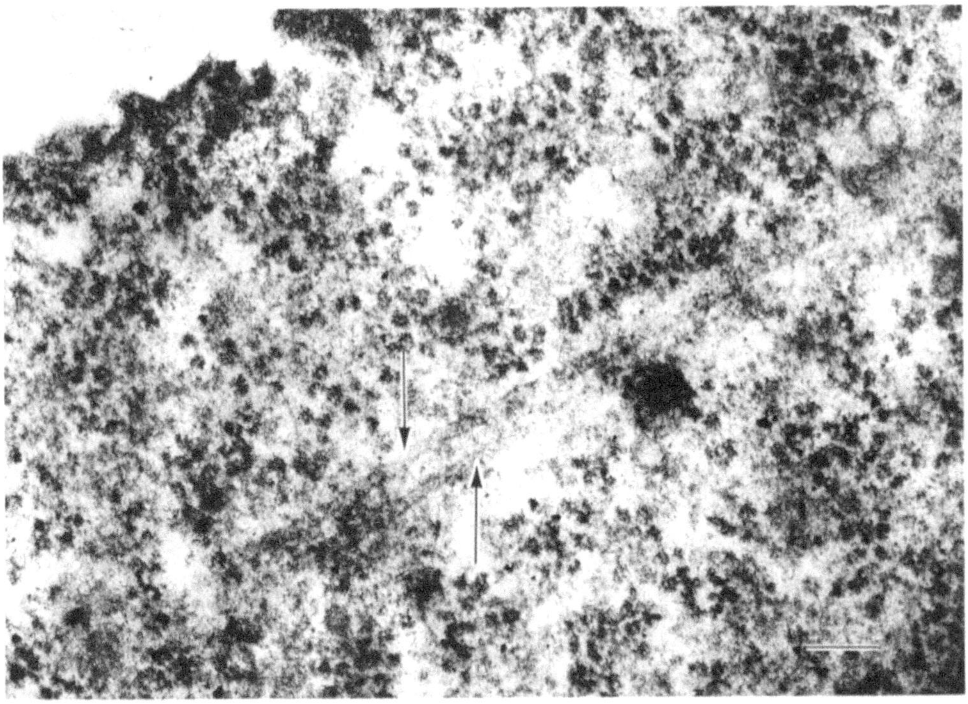

FIGURE 9. Electron micrograph of an ultrathin section through the central region of the primary pole of the mutant cdc 12-112. The plane of sectioning passes through two cytoplasmic microtubules (arrows) oriented in parallel to the axis of cell elongation. Bar equals 0.1 μm.

appears to be related to the attainment of a critical cell size or length rather than to particular timing in the cell cycle.[12] An examination of the systems of microtubules in these mutants would be one way to probe the function of these organelles.

V. CONCLUSIONS

Symmetrical cell division is a fundamental biological event; *Schiz. pombe* is one of only a few model systems for investigating the genetic control of this process. Cytokinetic and division-related cdc mutants of *Schiz. pombe* are usually stable; a reasonable proportion remains viable; and they are blocked synchronously at the restrictive temperature. It is hoped that a more representative range of cytokinetic mutants will be obtained in the near future and that these will facilitate manipulation of cytokinesis and an examination of its regulation. Some of this work might suggest other applications to higher cells.[47] Research on *Schiz. pombe* is aided by facile distinction of the lateral cell wall and the septum by primulin staining.[16,31]

The analysis of cytokinetic mutants of *Schiz. pombe* has led to the visualization of two classes of cytoplasmic cytoskeletal elements.[22] This allows comparisons between fission yeasts and animal cells because induction of cytokinesis in fungal and animal cells shares a number of features.[26] Our interpretation of the behavior of late septation and multiseptate mutants of *Schiz. pombe* has been influenced by data on animal cytokinesis,[20,30,33,35] and this is expected to continue. The animal systems do have the advantage of improved visualization of division-related cytoplasmic structures because microscopists do not have to contend with a cell wall in those organisms (see Volume II, Chapter 8). However, the expressive cell wall topography of *Schiz. pombe*, which constitutes a system of natural markers, will continue to be of considerable attraction for further research in this area.

ACKNOWLEDGMENTS

This communication is dedicated to Professor Manfred Girbardt on the occasion of his 60th birthday. I thank Paul Nurse for access to unpublished results and for stimulating discussions.

REFERENCES

1. **Streiblová, E. and Beran, K.**, Demonstration of yeast scars by fluorescence microscopy, *Exp. Cell Res.*, 30, 603, 1963.
2. **Streiblová, E. and Beran, K.**, Types of multiplication scars in yeast, demonstrated by fluorescence microscopy, *Folia Microbiol. (Prague)*, 8, 221, 1963.
3. **Streiblová, E. and Beran, K.**, On the question of the vegetative reproduction in apiculate yeasts, *Folia Microbiol. (Prague)*, 10, 352, 1965.
4. **Streiblová, E., Málek, I., and Beran, K.**, Structural changes in the cell wall of *Schizosaccharomyces pombe* during cell division, *J. Bacteriol.*, 91, 428, 1966.
5. **Streiblová, E.**, Cell division in yeasts, in *Recent Advances in Microbiology*, Pérez-Miravete A. and Peláez, D., Eds., Associacion Mexicana de Microbiologia, Mexico, DF, 1970.
6. **Streiblová, E. and Wolf, A.**, Role of the cell wall topography in conjugation of *Schizosaccharomyces pombe, Can. J. Microbiol.*, 21, 1399, 1975.
7. **Mitchison, J. M.**, The growth of single cells. I. *Schizosaccharomyces pombe, Exp. Cell Res.*, 13, 244, 1957.
8. **Johnson, B. F.**, Autoradiographic analysis of regional cell wall growth of yeasts, *Exp. Cell Res.*, 39, 613, 1965.
9. **Streiblová, E. and Wolf, A.**, Cell wall growth during the cell cycle of *Schizosaccharomyces pombe, Z. Allg. Mikrobiol.*, 12, 673, 1972.
10. **May, I. W.**, Sites of cell-wall extension demonstrated by the use of fluorescent antibody, *Exp. Cell Res.*, 27, 170, 1963.
11. **Johnson, B. F., Calleja, G. B., Boisclair, I., and Yoo, B. Y.**, Cell division in yeasts. III. The biased asymmetric location of the septum in the fission yeast cell *Schizosaccharomyces pombe, Exp. Cell Res.*, 123, 253, 1979.
12. **Nurse, P.**, personal communication, 1980.
13. **Johnson, B. F., Yoo, B. Y., and Calleja, G. B.**, Cell division in yeasts. II. Template control of cell plate biogenesis in *Schizosaccharomyces pombe*, in *Cell Cycle Controls,* Padilla, G.M., Cameron, I.L., and Zimmerman, A., Eds., Academic Press, New York, 1974.
14. **Oulevey, N., Deshuses, J., and Turian, G.**, Étude de la zone septale de *Schizosaccharomyces pombe* en division á ses étapes successives, *Protoplasma*, 70, 217, 1970.
15. **Johnson, B. F., Yoo, B. Y., and Calleja, G. B.**, Cell division in yeasts: Movement of organelles associated with cell plate growth of *Schizosaccharomyces pombe, J. Bacteriol.*, 115, 358, 1973.
16. **Nurse, P., Thuriaux, P., and Nasmyth, K.**, Genetic control of the cell division cycle in the fission yeast *Schizosaccharomyces pombe, Mol. Gen. Genet.*, 146, 167, 1976.
17. **Bonatti, S., Simili, M., and Abbondandolo, A.**, Isolation of temperature-sensitive mutants of *Schizosaccharomyces pombe, J. Bacteriol.*, 109, 484, 1972.
18. **Minet, M., Nurse, P., Thuriaux, P., and Mitchison, J. M.**, Uncontrolled septation in a cell division cycle mutant of the fission yeast *Schizosaccharomyces pombe, J. Bacteriol.*, 137, 440, 1979.
19. **Johnson, B. F. and Lu, C.**, Morphometric analysis of yeast cells. IV. Increase of the cylindrical diameter of *Schizosaccharomyces pombe* during the cell cycle, *Exp. Cell Res.*, 95, 154, 1975.
20. **Rappaport, R.**, Geometrical relations of the cleavage stimulus in invertebrate eggs, *J. Theoret. Biol.*, 9, 51, 1965.
21. **Nurse, P.**, unpublished results, 1980.
22. **Streiblová, E. and Girbardt, M.**, Microfilaments and cytoplasmic microtubules in cell-division cycle mutants of *Schizosaccharomyces pombe, Can. J. Microbiol.*, 26, 250, 1980.

23. **Schroeder, T. E.,** The contractile ring. I. Fine structure of dividing mammalian (HeLa) cell and the effect of cytochalasin B, *Z. Zellforsch.*, 109, 431, 1970.
24. **Girbardt, M.,** Ultrastructure of normal and affected wall formation in fungi, *Abstr. 12th Int. Bot. Congr.* Leningrad, 1975.
25. **Girbardt, M.,** Fungal cytokinesis—facts and problems, *Abstr. 2nd Int. Mycol. Congr.*, Bigelow, M. E. and Simmons E. G., Eds., Hamilton I. Newel, Tampa, 1977.
26. **Girbardt, M.,** A microfilamentous septal belt (FSB) during induction of cytokinesis in *Trametes versicolor, Exp. Mycol.*, 3, 215, 1979.
27. **Girbardt, M.,** Untersuchungen über die Querwandbildung bei Basidiomyceten, *Ber. Dtsch. Bot. Ges.*, 68, 423, 1955.
28. **Girbardt, M.,** Experimentelle Zellkern-Verlagerung in Pilz-hyphen, *Protoplasma*, 50, 483, 1959.
29. **Patton, A. M. and Marchant, R.,** An ultrastructural study of septal development in hyphae of *Polyporus biennis, Arch. Microbiol.*, 118, 271, 1978.
30. **Schroeder, T. E.,** Dynamics of the contractile ring, in *Molecules and Cell Movement,* Inoue, S. and Stephens, R. E., Eds., Raven Press, New York, 1975.
31. **Streiblová, E.,** Patterns of cell wall growth in cell-cycle mutants of *Schizosaccharomyces pombe. Abstr. 2nd Int. Mycol. Congr.*, Bigelow, H. E. and Simmons, E. G., Eds., Hamilton I. Newel, Tampa, 1977.
32. **Nurse, P.,** personal communication, 1980.
33. **Wolpert, L.,** Mechanics and mechanism of cleavage, *Int. Rev. Cytol.*, 10, 163, 1960.
34. **Hiramoto, Y.,** Analysis of cleavage stimulus by means of micromanipulation of sea urchin eggs, *Exp. Cell Res.*, 68, 291, 1971.
35. **Rappaport, R.,** Establishment and organization of the cleavage mechanism, in *Molecules and Cell Movement,* Inoue, S. and Stephens, R. E., Eds., Raven Press, New York, 1975.
36. **Arnold, J. M.,** Cytokinesis in animal cells: new answers to old questions, in *The Cell Surface in Animal Embryogenesis and Development,* Poste, G. and Nicolson, G. L., Eds., Elsevier North-Holland, Amsterdam, 1976.
37. **Johnson, B. F.,** Lysis of yeast cell walls induced by 2-deoxyglucose at their sites of glucan synthesis. *J. Bacteriol.*, 95, 1169, 1968a.
38. **Barras, D. R.,** A glucan endohydrolase from *Schizosaccharomyces pombe* and its role in cell wall growth, *Antonie van Leeuwenhoek; J. Microbiol. Serol.*, 38, 65, 1972.
39. **Johnson, B. F.,** Dissolution of yeast glucan induced by 2-deoxyglucose, *Exp. Cell Res.*, 50, 692, 1968b.
40. **Poole, R. K. and Lloyd, D.,** Effect of 2-deoxy-D-glucose on growth and cell walls of *Schizosaccharomyces pombe* 972 h^-, *Arch. Microbiol.*, 88, 257, 1973.
41. **Svoboda, A. and Smith, D. G.,** Inhibitory effect of 2-deoxyglucose on cell wall synthesis in cells and protoplasts of *Schizosaccharomyces pombe, Z. Allg. Mikrobiol.*, 12, 685, 1972.
42. **Biely, P., Farkaš, V., and Krátký, Z.,** Some aspects of wall formation in *Schizosaccharomyces pombe.* A study with 2-deoxyglucose, *Biologia (Bratislava)*, 29, 919, 1974.
43. **Mitchison, J. M.,** *The Biology of the Cell Cycle,* Cambridge University Press, London, 1971.
44. **Hereward, F. V.,** Cytoplasmic microtubules in yeast, *Planta*, 117, 355, 1974.
45. **Thuriaux, P., Nurse, P., and Carter, B.,** Mutants altered in the control of coordination of cell division with cell growth in the fission yeast *Schizosaccharomyces pombe. Mol. Gen. Genet.*, 161, 215, 1978.
46. **Nasmyth, K., Nurse, P., and Fraser, R. S. S.,** The effect of cell mass on the cell cycle timing and duration of S-phase in fission yeast, *J. Cell Sci.*, 39, 215, 1979.
47. **Pardee, A. B., Dubrow, R., Hamlin, J. L., and Kletzien, R. F.,** Animal cell cycle, *Ann. Rev. Biochem.*, 47, 715, 1978.

Chapter 5

PROTOPLASTS

W. N. Arnold

TABLE OF CONTENTS

I. INTRODUCTION

A protoplast is derived from a cell by complete removal of the cell wall and, concomitantly, all features external to the cell wall. Components of the periplasmic space are incidentally surrendered to the medium. Accordingly, a protoplast is bounded by the plasma membrane, but is denuded of all other components of the cell envelope. The osmotic sensitivity of protoplasts necessitates support by an external, non-penetrating solute to match the osmolality of the protoplasm. This requirement attends the actual generation of protoplasts as well as their sustenance.

En route to protoplasts, intermediary structures appear. Because the cell wall is partially removed, these derivative structures are already much less stable than intact cells towards changes in osmotic potential. The long-standing term for osmotically vulnerable entities of this type is "spheroplast". Some authors use the term "prospheroplast" to signify an even earlier stage, i.e., an osmotically sensitive unit with enough cell wall to prevent shape-change. This follows from the observation that protoplasts are invariably spherical, even when they are derived from the pyramidal cells of *Trigonopsis variabilis.* However, practicality of the term "prospheroplast" is questionable because the cells of some species or strains are inherently fairly spherical and no shape-change can be observed during protoplasting.

Distinction between spheroplast and protoplast ultimately rests on examination at the electron microscope level. Some investigators are apparently put off by the occasional difficulty in removing all cell wall remnants and, consequently, have fallen into the habit of calling all lysable preparations spheroplasts even when electron micrographs indicate protoplasts. On the other hand, to speak of "naked" protoplasts may be dramatic, but "naked" is redundant.

II. CELL WALL DISSOLUTION

Removal of the cell wall is the means to protoplast generation and in that one respect it matters little about the mechanisms involved. Therefore, rather crude mixtures of hydrolytic enzymes have been widely employed. Because glucans play the major role in cell wall mechanics of yeasts, all protoplasting agents have some glucanase component. These enzymes are able to attack β (1→3), β (1→6), or both types of linkage. The relative success achieved in generating protoplasts from different species and strains (or even the same strain under different growth conditions) is undoubtedly due to the relative proportion of the different glucan components in the walls and, perhaps more importantly, their accessibilities to the particular glucanases in the protoplasting fluid. Accessibility is apparently governed by the porosity of the overall fabric and is thought to be also influenced by interactions among the glucans and other polymers within the cell wall. Reducing agents, for example, play some role in modifying the accessibility of structural glucans to external glucanases.

By far the most popular protoplasting agent to date is the gastric juice of the snail, *Helix pomatia.* Snails can be dissected according to the description of Keilin[1] and the juices of the digestive track may be isolated according to Myers and Northcote.[2] But most workers now use commercially available preparations:

"Type H-2, Crude solution from *H. pomatia*"
 Sigma Chemical
 Saint Louis, Miss.
"Glusulase"
 Endo Laboratories
 Garden City, N.Y.

"Digestive juice of *H. pomatia*"
Industrie Biologique Francaise
Gennevilliers, France

Some commercial snail enzyme preparations contain an organic mercurical as an antiseptic. Because this compound may interfere with some yeast enzymes, it should be inactivated by sulfhydryl addition, or removed by percolating the crude mixture over cysteine-agarose (i.e., cysteine attached through the amino group to beaded agarose).

There is also insoluble matter in commercial snail juice. We routinely dilute 1 m*l* of concentrate with 9 m*l* of a suitable buffer containing osmotic fortification (see below) and 10 m*M* dithiothreitol. After centrifugation at 20,000 × g for 15 min, a greenish-brown pellet (apparently indicative of the snail's last meal) is visible. The clear amber supernate is used for protoplasting yeast cells.

Snail digestive juice contains 30 or more enzymes including at least 20 carbohydrases.[2,3] Because the juice contains small amounts of sucrase, maltase, and lactase, for example, as well as somewhat larger amounts of acid phosphatase, any studies relating to periplasmic enzymes in yeast or their secretion need to be adequately controlled. Some peptidases and proteinases are present, but are judged to be minor activities in sympathy with the snail's herbivorous nature. Yeast mannan is attacked to some extent, but β-linked mannans are not.[2] The major activity of interest in the present context is that which can be monitored by laminarin dissolution.[2,3] Here again a multiplicity of β-glucanases is suggested. An exo-(1→3)-β-glucanase has been considerably purified, has a pH optimum of 4.3, an estimated molecular weight of 82,000, but very little activity against laminarin at pH 6.5 and very little effect on yeast glucan.[4] A β-glucanase with endo-action pattern and much lower molecular weight[4] is apparently responsible for protoplast formation, but to the best of the author's knowledge has never been purified. Gel-permeation chromatography indicated four major protein-containing fractions in snail juice.[5] One fraction was responsible for protoplast generation from *Saccharomyces cerevisiae* or *Sacch. uvarum*, and there was some indication that at least two enzymatic components were involved. This followed from the inactivity of the "tails" of this peak when tested individually, in contrast to their activity in combination.[5] There has been no follow-up on this suggestion. Yeast mannan was not attacked by the salient fractions and proteinases were once again not implicated.

Another useful preparation is marketed under the name Zymolyase® by Kirin Brewery, Takasaka, Gumma Pref., Japan. This endo-(1→3)-β-glucanase is prepared from culture filtrates of *Arthrobacter luteus* and is said to be free of β-(1→2), β(1→4), and β(1→6) glucanases.[6] The enzyme has an estimated molecular weight of 21,000 and an optimum pH between 7.0 and 7.5 for attacking viable yeast cells.[7] Unfortunately, the commercial product still contains variable amounts of mannanase, proteinase, and acid phosphatase (Kirin Brewery, product bulletin). Some *Sacch. cerevisiae* strains (in particular NCYC 366) are easily protoplasted with zymolase.[8] It is too early to make a categorical statement, but the available reports suggest that snail gut enzymes will work on a wider range of yeast species.

Several other sources of β glucanases have been employed in protoplasting yeasts. Some of these may have appeal for a specific application, in which case a larger compilation[9] should be consulted. An enzyme preparation from *Oerskovia xanthineolytica* is active against some *Sacch. cerevisiae* strains[10] and is now commercially available (Teichozyme®, Worthington Biochemical Corporation, Freehold, N.J.). Teichozyme® is not recommended for generating protoplasts from *Sacch. rouxii.*[11]

There is a superabundance of methods in the literature for protoplast generation. For *Sacch. cerevisiae*, a slight modification of the method of Cabib[12] is recommended, i.e., use 1.0*M* sorbitol as osmoticum, rather than 10% w/v mannitol. Sorbitol is much more soluble in water than is mannitol. This is of some convenience in making higher strength solutions for fortifying snail enzyme and other components in the mixture. The review by Villanueva and Acha[9] is a valuable secondary reference.

III. CHOICE OF OSMOTICUM

The two osmotica in vogue are sorbitol and potassium chloride. The historic method of Eddy and Williamson[13] recommended 0.55*M* rhamnose, but that sugar is about 34 times as expensive as sorbitol. Many other nontransported sugars have been suggested at concentrations ranging from 0.4 to 1.2 *M*, but 1.0 *M* sorbitol should be the best starting point in exploring a new species. For osmo-tolerant species (e.g., *Sacch. rouxii* and *Sacch. bisporus*) at least 2.0 *M* sorbitol is required, although potassium chloride is probably better.[11,14]

Utile salts for osmotic fortification include sodium and potassium chlorides and magnesium sulfate. The recommended salt is potassium chloride and concentrations in the range of 0.3 to 1.0 *M* are most common in the literature. Arnold and Lacy[15] rationalized the vogue concentration of 0.6 *M* KCl for *Sacch. cerevisiae* by demonstrating that this is close to the osmolality which elicits incipient plasmolysis. All other formulations are empirical. There is an indication that lower osmolalities (i.e., corresponding to 0.3 *M* KCl) are required for *Schizosaccharomyces pombe*,[16] and 2.0 *M* KCl is recommended for osmotolerant species.[11,14]

According to Willsky,[16a] sorbitol inhibits plasma membrane ATPase but KCl does not. Different methods of isolation may explain some of the discrepancies concerning that enzyme (see Volume II, Chapter 1). Matern[16b] has shown that 0.6 *M* KCl supported protoplasts display catabolite inactivation of phosphoenolpyruvate carboxykinase at nearly the same rate as that of intact cells. On the other hand 0.9 *M* mannitol does not allow catabolite inactivation unless certain ions are also added. Hutchison and Hartwell[17] claim that 5 m*M* NaCl, KCl, $MgCl_2$, or $(NH_4)_2SO_4$ significantly inhibits the actual production of protoplasts in the presence of 1.0*M* sorbitol; a curious result that has received no subsequent attention.

In the case of higher plant protoplasts, the type of osmoticum, e.g., mannitol vs. KCl, can affect the nature of the protoplast surface as evidenced by differences in rupture mechanisms attending dilution. Thus, mannitol supported protoplasts break into many small particles, whereas salt supported protoplasts yield relatively intact ghosts.[18] In the same vein, the susceptibility of alditol-supported yeast protoplasts to ribonuclease (and other basic proteins) is well known, whereas the protection by ions is apparently related to interactions among negatively charged groups on the plasma membrane.[19] These differences point up potential discrepancies in the behavior of protoplasts prepared under the two different types of osmotica.

IV. SULFHYDRYL EFFECTS

Starting with the first report by Eddy and Williamson,[13] it has been noted that some strains of *Sacch. cerevisiae* are refractory to protoplasting. Morever, most strains become more refractory in stationary phase. Schlenk's laboratory found that *Candida utilis* cells cultivated in the presence of methionine are more susceptible to the action of snail gut enzyme than ordinary cells.[20] The next step was to show that reducing agents will increase the susceptibility of normally refractory cells.[21,22] The positive effect of cysteine, dithiothreitol, thioglycollate, 2-mercaptoethanol, or 2-mercaptoethanolamine is elicted in sugar or salt fortified media. Probably 10 m*M* dithiothreitol will serve best in the exploration of any new species.

Bacon et al.[73] found that isolated cell walls of bakers' yeast required the presence of a reducing agent for digestion by the β-glucanases of *Cytophaga johnsonii*. The reducing agents clearly exert their effect on the cell wall rather than the β-glucanases.[22,23] Killick[24] attempted to extend the result of Svihla et al.[20] by seeking a meaningful chemical change in the walls of cells grown in high sulfur media. Indeed much higher proportions of protein and lipid were found in the cell walls of supplemented *Sacch. cerevisiae*, but apparently not in those of *C. utilis.*[24] Bacon et al.[23] found that autoclaving cell walls seemingly substitutes for treatment with reducing agent as far as their susceptibility is concerned. There is no indication that the positive effect of reduction is reversible. While studying the ethylacetate treated model (see Volume II, Chapter 7), this was attempted by bubbling oxygen through a suspension of reduced cell ghosts, but no decrease in subsequent susceptibility to β-glucanases was found.[25]

Schwencke et al.[26] claim that a formulation including dithiothreitol, ethylenediaminetetraacetic acid, a pH 8.9 buffer, and 1 mg/m*l* pronase is a good pretreatment solution to aid in the production of protoplasts from stationary phase cells of *Sacch. cerevisiae, Endomycopsis fibuligera,* or *C. tropicalis.* Unfortunately, no data on the same aged cells was presented to show the effects of withdrawing one item at a time from the complex mixture. (Others have recommended the use of ethylenediaminetetraacetic acid to prevent clumping;[22] we find that a higher initial concentration of snail enzymes ameliorates that condition). Without evidence to the contrary one would fear that pronase might be deleterious to the plasma membrane, and the choice of pH 8.9 is definitely outside the range of kind pH for yeasts. Furthermore, the yields of stable protoplasts from some strains of *Sacch. cerevisiae* were as low as 25% in some cases, so the method of Schwencke et al.[26] cannot be recommended generally. Their results do, however, add to the growing body of results that implicates a protein component interacting with the cell wall glucans.

Saccharomycopsis guttulata cell walls contain more protein and proportionally less mannan than do walls of *Sacch. cerevisiae.*[27] *Sacch. cerevisiae* walls are attacked by a β(1→3) glucanase from *Bacillus circulans*[28] only after addition of 2-mercaptoethanol. The walls of *S. gluttulata* are susceptible to this enzyme but are only slightly more so in the presence of 2-mercaptoethanol. On the other hand, pronase pretreatment increases the susceptibility of *S. guttulata* walls to β-glucanases.[27] Although this does not divorce the sulfhydryl effect from the protein component, it does suggest that the effect is a complex one or that the cell wall proteins vary in their cystine content among different species.

So, although the use of sulfhydryl agents is firmly entrenched as an aid to protoplasting, we still do not know exactly what reducing agents do to cell walls to make them more vulnerable. Hutchison and Hartwell[17] showed that a 10 min pretreatment with 1 m*M* dithiothreitol (or a variety of other reducing agents) markedly inhibited RNA synthesis in intact cells of *Sacch. cerevisiae*. For this reason they caution the use of sulfhydryl agents if protoplasts are intended for studies on macromolecular syntheses.

V. COURSE OF PROTOPLAST FORMATION

The dissolution of the cell wall is never completely uniform over the external surface. This seems to indicate that the loci of primary attack have specific chemical or physical differences from the bulk wall structure. But whether these loci are distributed in random or organized fashion is not known.

Eddy and Williamson[13] observed the extrusion of protoplasts through holes in the cell walls, and this is a common observation that has been documented by light and electron microscopy.[29] Pringle et al.[8] presented scanning electron micrographs of *Sacch.*

cerevisiae in various stages of cell wall dissolution. The authors remarked on the absence of protoplast extrusion, but rather noted the development of large fissures followed by sloughing off of large pieces of cell wall. They attribute the different mechanism to the relative susceptibility of strain NCYC 366. However, Arnold believes that whether extrusion is displayed or not, depends upon the osmolality of the medium during protoplasting and how closely it matches the initial osmolality of the protoplasm. For example, Eddy and Williamson[13] employed 0.55 *M* rhamnose, whereas, Pringle et al.[8] used 1.2 *M* sorbitol. If hypertonic solutions are employed, it would seem physically impossible to get extrusion. If the medium is isotonic with the average cell, then both mechanisms might be displayed depending upon the range of cytoplasmic osmolalities possessed by the particular sample.

Protoplasts from a variety of *Saccharomyces* and *Candida* species retain furrows in the plasma membrane under a range of osmolalities in the support medium. The pattern of these furrows seems to vary among strains of *Sacch. cerevisiae*[8,30] and among other species.[30] The interpretation of scanning electron micrographs equates these furrows with invaginations in the plasma membrane previously reported by transmission electron microscopy of whole cells and of protoplasts in thin section. The present notion is that these furrows are related to expansion of the plasma membrane during growth and are not due to crenation under hypertonicity as some had previously proposed.

Small balls of unknown chemistry, but supposedly of wall origin, appear on the surface of protoplasts during the course of cell wall digestion. These globules appear during zymolase or snail enzyme treatments.[8,30] They may disappear after prolonged incubation. On the other hand, the bud scars resist digestion of their chitin content and are sloughed off as characteristic annuli.[31]

There is no substitute for electron microscopy in making an estimate on the relative prevalence of protoplasts and spheroplasts in digestion mixtures. However, after preliminary exploration of conditions, electron microscopy is neither practical nor essential for routine preparations. Rather, the time course is typically monitored by estimating osmotic sensitivity (protoplasts plus spheroplasts). At zero time and defined periods thereafter, two small portions are taken. One is diluted 100-fold with distilled water, the other with osmotically fortified medium, and concentrations are estimated under the light microscope with a hemacytometer. It is also convenient to establish a standard curve relating turbidity (e.g., Klett value) to the pecentage of derivative structures that are osmotically vulnerable.

When the production of lysable units is judged to have reached a maximal value, the protoplasts or spheroplasts are harvested by centrifugation and washed by resuspension in osmotically fortified medium. Protoplasts are apparently not hurt by centrifugation (even at 100,000 × g).[32] However, they are vulnerable to shearing forces which develop if resuspension of pellets is done too vigorously. Accordingly, it is recommended that a fine glass rod be used to mingle each pellet with a minimum of resuspension fluid prior to gentle vortexing to final dilution. Depending on the application, it may be more desirable to employ centrifugation on a Ficoll® gradient to separate protoplasts from cell wall debris, and from any cells that have escaped protoplasting. By judicious choice of densities, the resultant band that is enriched in protoplasts is cleansed of snail enzymes and, because of the high-density cushion, resuspension is facilitated and lysis is minimal.[11]

VI. IS A PROTOPLAST A "CELL WITHOUT A WALL"?

It is well known that yeast protoplasts may show a gamut of vital signs and, under the appropriate conditions, exhibit various aspects of morphogenesis (see Volume II, Chapter 6). Therefore, the short answer to the above question is that protoplasts are

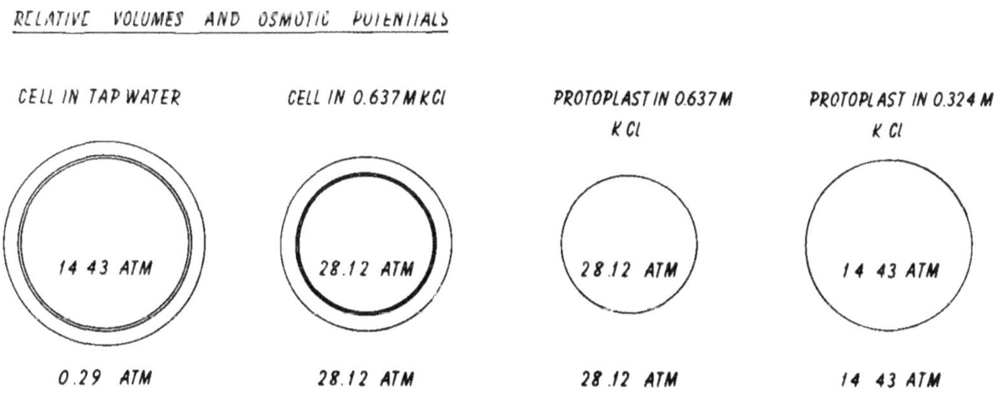

FIGURE 1 Diagrammatic representation of yeast cell and protoplast cross sections which reflect relative sizes in different media. Cytoplasmic and medium osmotic potentials are indicated in each case. Calculations are given in the text.

probably good models for cells. However, it is not trivial to wonder how much trauma is sustained by the protoplasm during protoplast formation. The possible influences of osmotically fortified media on protoplast physiology and biochemistry are also worth considering. A few simple calculations help to make the point.

Conway and Armstrong[33] used freezing-point depression measurements to calculate solute concentrations in a bakers' yeast equilibrated with tap water. For resting yeast, the quoted internal concentration of 0.590 osmol kg^{-1} equates to 14.43 atm osmotic potential at 25°C. The tap water in Dublin is not described, but in Kansas City it is about 0.012 osmol kg^{-1} (or 0.29 atm). This means that the "average" yeast cell, as represented in Figure 1, has a wall pressure (exerted on the protoplasm) of (14.43 – 0.29) viz., 14.14 atm.

Now Red Star bakers' yeast in the hands of Arnold and Lacy[15] showed incipient plasmolysis in a medium of 1.15 osmol kg^{-1} (i.e., 28.12 atm). This means that for an average cell equilibrated with 0.637 *M* KCl, the wall pressure is zero and the internal osmotic potential is 28.12 atm (see Figure 1). If an average cell in tap water is transposed into (and equilibrated with) 0.637 *M* KCl, then the protoplasm must shrink to 51.3% of its initial value. For spheres the final diameter would be 80.1% of the initial value. The same is true for protoplasts (Figure 1) prepared in 0.637 *M* KCl (or 1.01 *M* alditol). The protoplasts could be returned to the size of the original protoplasm (Figure 1) by transferring them to a 0.590 osmol kg^{-1} medium, i.e., 0.324 *M* KCl (or 0.544 *M* alditol). This, of course, assumes an elastic plasma membrane and no membrane pressure. The author is not aware of any such intentional manoeuvre of isolated protoplasts into a less concentrated medium so that they might approximate the size of the protoplasm in the original cells. The rate of synthesis of β-fructofuranosidase by *Sacch. cerevisiae* protoplasts is increased about 40% when supported by 0.4 *M* KCl as opposed to the 0.6 *M* KCl,[33a] and this is in sympathy with the above argument.

The above argument does not consider possible losses of internal solutes during protoplasting, but rather treats the protoplast as a perfect osmometer. There is also the possibility that the plasma membrane may be physically modified during protoplast generation. When higher plant cells are converted into protoplasts, at least 25% of the plasma membrane can be lost, and ends up as vesicles in the cytoplasm, probably by a pinocytosis-like process.[34] We have no evidence for this type of internalization in yeast, but have wondered about externalization in connection with the origin of periplasmic bodies (vesicles which appear in the periplasmic space) as discussed in Volume II, Chapters 1 and 8.

Hutchison and Hartwell[17] reported very reasonable rates of RNA, DNA, and protein syntheses by *Sacch. cerevisiae* protoplasts. However, their summary comparisons were made with values for whole cells in similarly fortified media (e.g., 1.0 *M* sorbitol) and it has apparently escaped subsequent reviewers that the rates for cells were also depressed by increasing the osmolality of the medium. Based on approximations from published graphs,[17] the author has estimated the 4 hr increment in RNA (apparently the most sensitive process of those mentioned above). Compared with unfortified medium, the rates were 82% and 43% for cells in 0.9 *M* and 1.3 *M* sorbitol, respectively. The companion rates for protoplasts were 38% and 25% (with no correction for losses during protoplasting).

In sum, it would seem that protoplasts have been, and will continue to be, useful models of cellular behavior. Likewise, the plasma membrane obtained from lysed protoplasts is a useful material for studying certain enzyme activities (see Volume II, Chapter 1 and the following section). Nevertheless, changes in the environment that attend cell wall dissolution are evident and it must be anticipated that some enzymatic processes will be more affected than others. Manipulation of the support medium to one that is isotonic with resting protoplasm (see above) might ameliorate some of these problems or highlight interesting mechanisms.

VII. LYSIS

Protoplasts have been successfully stored in 0.6 *M* KCl at 25°C for a week and at 4°C for over 2 months.[20] These are records and are not recommended practice, although several investigators report convenient storage for several days at 4°C. Cabib[12] recommends chlortetracycline (50 μg/m*l*) to prevent contamination by bacteria. One of the most thorough studies of protoplast stability is that of Svihla et al.[20] They used snail gut enzymes to generate protoplasts from *C. utilis* as well as *Sacch. cerevisiae* cells. Digestion medium contained 0.6 *M* KCl and 0.04 *M* potassium phosphate buffer (pH 6.0). Subsequent stability studies were based on microscopic examination and on release of isotopically labeled cytoplasmic compounds. When *C. utilis* protoplasts were rapidly diluted (100-fold) in 0.5 to 1.0 *M* KCl, there was no significant degree of lysis. With lower concentrations, a graded response to 100% lysis at 0.006 *M* KCl was observed. Incidentally, in the region of 0.3 *M* KCl (compare this concentration with theoretical argument in Section VI) the degree of lysis recorded was about 50%. But the speed of dilution is an important parameter here; there is every indication that much lower osmolalities are tolerated by yeast and bacterial protoplasts if they are slowly adjusted (e.g., by dialysis).[9]

Lysis by dilution is the most common means to prepare plasma membranes from protoplasts. An innovation due to Scarborough,[35] wherein fungal plasma membranes are reinforced with concanavalin A, greatly facilitates the production of large pieces of yeast plasma membrane rather than reformed vesicles. Various applications in enzyme studies are mentioned in Volume II, Chapter 1. The role of lipid constituents in plasma membrane resistance, and the manifestations of altering the lipid composition by diet or temperature manipulation are discussed in Volume I, Chapter 7.

Mechanical disruption has already been mentioned in the context of undesirable lysis attending resuspension of pellets. For intentional disruption, simultaneous dilution and imposition of shearing forces may be preferred. Probably the most controllable energy input is via ultrasonic oscillation. Protoplasts are rapidly lysed by 1 to 2 min exposures, whereas intact cells are very resistant[20] (and see also Volume I, Chapter 3).

Yeast protoplasts are also susceptible to UV radiation in the region of 260 nm. Doses of the order of 6×10^3 J m^{-2} cause lysis of protoplasts rather quickly, whereas vacuoles require about three times that dose.[20]

Protoplasts may lyse as a result of their own metabolic activities. Abrams[36] showed that protoplasts of *Streptococcus faecalis* undergo a metabolic swelling in the presence of glucose and K^+. The degree of swelling reached a maximum corresponding to the complete conversion of glucose to lactic acid and thereafter the protoplasts shrunk, i.e., the process was completely reversible. An excessive amount of glucose (approximately 20 m*M*) caused lysis in the presence of 0.3 *M* sucrose as osmoticum. Higher concentrations of the osmoticum allowed bacterial protoplasts to withstand higher initial concentrations of glucose. Most of the basic work on metabolic lysis by yeast protoplasts has come from Indge.[37-39] Protoplasts resting in 0.55 *M* mannitol lyse upon the addition of 30 m*M* glucose.[38] Osmotic susceptibility is accentuated by inclusion of chelating agents,[37,38] but protoplasts are somewhat protected by metabolic inhibitors and alkali metal ions.[38] This method has been used to isolate vacuoles[39] as well as to yield starting material for plasma membrane purification.[40] Because vacuole breakage is avoided during metabolic lysis of protoplasts, one can save the plasma membrane from exposure to some potentially deleterious proteolytic enzymes.

VIII. SOME APPLICATIONS

Within certain restraints yeast protoplasts are very convenient biological material for a variety of biochemical and biophysical investigations. Some of these have been mentioned in passing.

The area of genetic engineering has blossomed and in the context of manipulation of yeasts there has been much interest in the insertion of nucleic acids from one species into another. All of these studies have been done with protoplasts. The subject of fusion between protoplasts and its prospects are addressed in Volume II, Chapter 6. From a practical point of view, there is immediate interest in incorporating the genetic information for amylase synthesis into brewing yeasts so that they might take full advantage of the carbohydrates in the wort. Because yeast species (particularly *Sacch. cerevisiae*) have been extensively studied in terms of genetics, and because of the availability of a great selection of defined strains, they are prime candidates for certain studies requiring a eukaryotic cell model.

The phenomenon of cell wall regeneration in yeast protoplasts (see Volume II, Chapter 6) provides another approach to some intriguing questions about the disposition of polymers within the cell envelope. The use of protoplasts to study secretion of enzymes normally destined for the periplasmic space has helped in localization of these glycoproteins and has lead to several working hypotheses on mechanism (see Volume II, Chapter 1).

Most of the studies on yeast protoplasts have involved *Sacch. cerevisiae* and a few other species. Protoplasts of yeast and yeast-like species that are pathogenic to man have not been studied in any depth, but *Cryptococcus neoformans*,[41] *Histoplasma capsulatum*,[42] *Blastomyces dermititidis*,[42] *Candida albicans*,[43] and *Torulopsis glabrata*[43] are all susceptible to protoplasting by snail gut enzymes. Domanski and Miller[44] claim that a chitinase from *Streptomyces griseus* helps for *C. albicans* but no follow-up has appeared on this. There is sufficient preliminary information on hand to indicate that informative studies with antimycotics could be carried out on protoplasts from pathogenic yeasts.

IX. CONCLUDING REMARKS

In many instances the main purpose in making protoplasts is to explore the plasma membrane and not have to contend with the cell wall. To this end, it is important not to damage the plasma membrane. Thus, it seems clear that proteolytic and lipolytic activities should be avoided in any protoplasting fluid. Enzymes from the snail digestive tract continue to be the most popular agents because they attack the widest range of yeast species. However, there is a need for more purified β-glucanase preparations and zymolase seems to fill the bill in some cases. Further attempts to fractionate the snail product might be fruitful. It is surprising that very little has appeared along that line since the gel-permeation chromatography studies of the late 1960s.

REFERENCES

1. **Keilin, J.,** Helicorubin and cytochrome h purification and physical properties, *Biochem, J.*, 64, 663, 1956.
2. **Myers, F. L., and Northcote, D. H.,** A survey of the enzymes from the gastro-intestinal tract of *Helix pomatia, J. Exp. Biol.*, 35, 639, 1958.
3. **Holden, M. and Tracey, M. V.,** A study of enzymes that can break down tobacco-leaf components. II. Digestive juice of *Helix* on defined substrates, *Biochem, J.*, 47, 407, 1950.
4. **Marshall, J. J. and Grand, R. J. A.,** Comparative studies on β-glucan hydrolases. Isolation and characterization of an exo (1→3) -β-glucanase from the snail, *Helix pomatia, Arch. Biochem. Biophys.*, 167, 165, 1975.
5. **Anderson, F. B. and Millbank, J. W.,** Protoplast formation and yeast cell wall structure. The action of the enzymes of the snail, *Helix pomatia, Biochem, J.*, 99, 682, 1966.
6. **Kitamura, K. and Yamamoto, Y.,** Purification and properties of an enzyme, Zymolyase, which lyses viable yeast cells, *Arch. Biochem. Biophys.*, 153, 403, 1972.
7. **Kitamura, K., Kaneko, T., and Yamamoto, Y.,** Lysis of viable yeast cells by enzymes of *Arthrobacter luteus*. II. Purification and properties of an enzyme, Zymolyase, which lyses viable yeast cells, *J. Gen. Appl. Microbiol.*, 20, 323, 1974.
8. **Pringle, A. T., Forsdyke, J., and Rose, A. H.,** Scanning electron microscope study of *Saccharomyces cerevisiae* spheroplast formation, *J. Bacteriol.*, 140, 289, 1979.
9. **Villanueva, J. R. and Acha, I. G.,** Production and use of fungal protoplasts, in *Methods in Microbiology,* Vol. 4, Booth, C., Ed., Academic Press, London, 1971, chap. 24.
10. **Mann, J. W., Heintz, C. E., and Macmillan, J. D.,** Yeast spheroplasts formed by cell-wall degrading enzymes from *Oerskovia* sp., *J. Bacteriol.*, 111, 821, 1972.
11. **Arnold, W. N. and Garrison, R. G.,** Isolation and characterization of protoplasts from *Saccharomyces rouxii, J. Bacteriol.*, 137, 1386, 1979.
12. **Cabib, E.,** Yeast spheroplasts, *Methods of Enzymol.*, 22, 120, 1971.
13. **Eddy, A. A. and Williamson, D. H.,** A method of isolating protoplasts from yeast, *Nature (London)*, 179, 1252, 1957.
14. **Weimberg, R. and Orton, W. L.,** Elution of acid phosphatases from the cell surface of *Saccharomyces mellis* by potassium chloride, *J. Bacteriol.*, 90, 82, 1965.
15. **Arnold, W. N. and Lacy, J. S.,** Permeability of the cell envelope and osmotic behavior in *Saccharomyces cerevisiae, J. Bacteriol.*, 131, 564, 1977.
16. **Svoboda, A., Farkâs, V., and Bauer, Š.,** Response of yeast protoplasts to 2-deoxyglucose, *Antonie van Leeuwenhoek, J., Microbiol. Serol.*, 35 (Suppl.) B11, 1969.

16a. **Willsky, R. G.,** personal communication.

16b. **Matern, H.,** Catabolite inactivation of phosphoenolpyruvate carboxykinase in spheroplasts from *Saccharomyces cerevisiae, Biochim. Biophys. Acta*, 586, 258, 1979.

17. **Hutchison, H. T. and Hartwell, L. H.,** Macromolecular synthesis in yeast spheroplasts, *J. Bacteriol.*, 94, 1697, 1967.
18. **Ruesink, A.,** Surface membrane properties of isolated protoplasts, *Colloq. Int. C.N.R.S.*, 212, 41, 1973.
19. **Schlenk, F. and Dainko, J. L.,** Action of ribonuclease preparations on viable yeast cells and spheroplasts, *J. Bacteriol.*, 89, 428, 1965.
20. **Svihla, G., Schlenk, F., and Dainko, J. L.,** Spheroplasts of the yeast *Candida utilis, J. Bacteriol.*, 82, 808, 1961.
21. **Davies, R. and Elvin, P. A.,** The effect of β-mercaptoethanol on release of invertase and formation of protoplasts of *Saccharomyces fragilis, Biochem, J.*, 93, 8P, 1964.

22. **Duell, E. A., Inoue, S., and Utter, M. F.**, Isolation and properties of intact mitochondria from spheroplasts of yeast, *J. Bacteriol.*, 88, 1762, 1964.
23. **Bacon, J. S. D., Milne, B. D., Taylor, I. F., and Webley, D. M.**, Features of the cell-wall structure of yeast revealed by the action of enzymes from a non-fruiting myxobacterium *(Cytophaga johnsonii), Biochem, J.*, 95, 28 C, 1965.
24. **Killick, K. A.**, Modification of the composition and structure of the yeast cell wall by culture in the presence of sulfur amino acids, *J. Bacteriol.*, 106, 931, 1971.
25. **Arnold, W. N.**, The structure of the yeast cell wall. Solubilization of a marker enzyme, β-fructofuranosidase by the autolytic enzyme system, *J. Biol. Chem.*, 247, 1161, 1972.
26. **Schwencke, J., Magana-Schwencke, N., and Laporte, J.**, Yeast protoplasts from stationary and starved cells: preparation ultrastructure and vacuolar development, *Ann. Microbiol.*, 128A, 3, 1977.
27. **Buecher, E. J. and Phaff, H. J.**, Cell wall composition of *Saccharomycopsis* (Schönning), *Proc. 2nd Int. Symp. Yeast Protoplasts*, Brno, 1968, 165.
28. **Tanaka, H. and Phaff, H. J.**, Enzymatic hydrolysis of yeast cell walls. I. Isolation of wall decomposing organisms and separation and purification of lytic enzymes, *J. Bacteriol.*, 89, 1570, 1965.
29. **Darling, S., Theilade, J., Birch-Andersen, A.**, Kinetic and morphological observations on *Saccharomyces cerevisiae* during spheroplast formation, *J. Bacteriol.*, 98, 797, 1969.
30. **Miegeville, M. and Morin, O.**, Nouvelle contribution de la microscopie electronique à balayage à l étude des protoplastes de levures, *C. R. Acad. Sci. Ser. D.*, 284, 1935, 1977.
31. **Bacon, J. S. D., Davidson, E. D., Jones, D., and Taylor, I. F.**, The location of chitin in the yeast cell wall, *Biochem, J.*, 101, 36C, 1966.
32. **Nakamura, K. D. and Schlenk, F.**, Stability of *Candida utilis* cells and spheroplasts toward gravitational forces, *J. Bacteriol.*, 116, 1472, 1973.
33. **Conway, E. J. and Armstrong, W. N.**, The total intracellular concentration of solutes in yeast and other plant cells and the distensibility of the plant cell wall, *Biochem, J.*, 81, 631, 1961.
33a. **Kuo, S. C. and Lampen, J. O.**, Osmotic regulation of invertase formation and secretion by protoplasts of *Saccharomyces, J. Bacteriol.*, 106, 183, 1971.
34. **Ruesink, A. W.**, Protoplasts of plant cells, *Methods Enzymol.*, 23A, 197, 1971.
35. **Scarborough, G. A.**, Isolation and characterization of *Neurospora crassa* plasma membranes, *J. Biol. Chem.*, 250, 1106, 1975.
36. **Abrams, A.**, Reversible metabolic swelling of bacterial protoplasts, *J. Biol. Chem.*, 234, 383, 1959.
37. **Indge, K. J.**, The effects of various anions and cations on the lysis of yeast protoplasts by osmotic shock, *J. Gen. Microbiol.*, 51, 425, 1968.
38. **Indge, K. J.**, Metabolic lysis of yeast protoplasts, *J. Gen. Microbiol.*, 51, 433, 1968.
39. **Indge, K. J.**, The isolation and properties of the yeast cell vacuole, *J. Gen. Microbiol.*, 51, 441, 1968.
40. **Cabib, E., Ulane, R., and Bowers, B.**, Yeast chitin synthetase. Separation of the zymogen from its activating factor and recovery of the latter in the vacuole fraction, *J. Biol. Chem.*, 248, 1451, 1973.
41. **Peterson, E. M., Hawley, R. J., and Calderone, R. A.**, An ultrastructural analysis of protoplast-spheroplast induction in *Cryptococcus neoformans, Can. J. Microbiol.*, 22, 1518, 1976.
42. **Berliner, M. D. and Reca, M. E.**, Protoplasts of systemic dimorphic fungal pathogens: *Histoplasma capsulatum* and *Blastomyces dermatitidis, Mycopath. Mycol. Appl.*, 37, 81, 1969.
43. **Brown, J. P.**, Susceptibility of the cell walls of some yeasts to lysis by enzymes of *Helix pomatia, Can. J. Microbiol.*, 17, 205, 1971.
44. **Domanski, R. E. and Miller, R. E.**, Use of a chitinase complex and β-(1→3)-glucanase for spheroplast production from *Candida albicans, J. Bacteriol.*, 96, 270, 1968.

Chapter 6

MORPHOGENESIS IN PROTOPLASTS

O. Nečas and A. Svoboda

TABLE OF CONTENTS

I. CELL WALL REGENERATION AND PROTOPLAST REVERSION

A. Introduction

The fact that protoplasts of green algae and molds are able to build new cell walls was recognized at the end of the last century (for review see Nečas[1]). At that time protoplasts were obtained by plasmolysis and mechanical dissection. These techniques could hardly be applied to yeasts. The successful large-scale preparation of yeast protoplasts was achieved first by controlled mechanical disruption of cells,[2] later by autolysis of cell walls,[3] and lately by enzymatic digestion of cell walls (see Volume II, Chapter 5). Regeneration of cell walls in yeast protoplasts, and reproductive ability of walled protoplasts, were first reported by Nečas in 1955 for *Saccharomyces cerevisiae.*[3,4] In these experiments protoplasts were cultivated on the surface of solid nutrient media. That method yielded a very low frequency of regeneration (only about 0.1%). Later the rate of regeneration was increased considerably by embedding protoplasts in gelled media.[5]

Once conditions for large-scale preparation of protoplasts and for high-frequency regeneration were at hand, detailed studies of protoplast regeneration began. In all yeast species studied to date, regeneration of cells follows the same pattern which includes three main events, i.e., growth of the protoplasts, generation of new cell walls, and reversion of protoplasts into cells.[1]

The conversion of yeast cells to protoplasts inhibits cytokinesis. Otherwise all synthetic processes continue. Protein, RNA, and DNA syntheses persist and all intracellular organelles, including nuclei, reproduce. Protoplast growth continues without cytokinesis until a new cell wall is assembled on the protoplast surface at which time the process of reversion is then initiated.

First signs of the new cell wall appear after a short-lag period. The length of the cell wall regeneration period differs considerably with the yeast species and particularly with conditions of protoplast cultivation. Wall regeneration is deemed to be finished when it reaches a stage of complexity that allows initiation of morphogenetic processes. This is the beginning of protoplast reversion and the hallmark is initiation of the mechanisms of cytokinesis.

Reversion is a complex process. Protoplasts generate a new cell wall and then give rise to new cells, revertants, with properties approaching those of the original cells. The morphology of protoplast reversion is species-specific. Moreover, it is strongly influenced by the previous history of the protoplast. The revertants often undergo several divisions before they can be judged to execute proper (gene-controlled) cell division morphology. Protoplast growth and cell wall regeneration occur simultaneously whereas reversion must be preceded by wall genesis. If suitable conditions for wall generation are not provided or if the generation is inhibited experimentally, the protoplasts continue to grow, but only for a limited period.[6,7] Under other specific circumstances described below protoplast reversion does not occur even in the presence of a new wall. Yeast protoplasts undergoing reversion to cells provide a unique model for studying many cellular processes. These include mechanisms of cell wall biogenesis and the role of the cell wall in morphogenetic processes such as cell shape determination, dimorphism, and sexual conjugation.

The fact that nuclear division is not in concert with protoplast growth offers an unusual opportunity to study the initiation of karyokinesis.[8] Also, cytokinesis is not commenced unless the regenerating cell wall attains a certain degree of complexity.[9] What are the biochemical and morphological characteristics that define this complexity? It is possible that attaining this stage of complexity as such enables the regenerated cell wall to participate actively in cytokinesis, or that the regenerated wall sends signals to other cell structures (notably the cytoskeleton) which then trigger cytokinesis.

On the other hand, the protoplast model has certain limitations. Cell wall genesis by protoplasts and reversion to cells must be regarded, *strictu senso*, as repair processes and controls may be minimal. Moreover, each living system when under stress may resort to special processes remote from the norm. Such repair processes, although coded for in the cell genome, are inactive during the normal physiological state. Thus, analogies between the normal and the repair processes must be made with caution.

B. Wall Regeneration

1. *General Pattern*

The cell wall genesis is a hierarchical process consisting of a set of sequential steps. As a working hypothesis one may consider the following pattern. First, various precursors are synthesized; these include glucans, mannoproteins, and chitin. Second, the precursors are transported to appropriate sites; this includes translocation through the plasma membrane. Third, the precursors are assembled as microfibrils and wall matrix material. And fourth, these components are assembled into the final biostructure which is recognized as a wall and cell morphogenesis can proceed.

Cell wall genesis by protoplasts starts from a situation which does not resemble the normal wall development in cells during vegetative growth in that there is no barrier to modify the flow of secreted wall precursors. It is now well established (and see also Volume II, Chapter 5) that the protoplast surface can be denuded of all original wall material which might anchor the precursors and thus act as a template giving epigenetic information for the three-dimensional assembly of wall precursors (for review see Nečas[9]). In this sense cell wall generation by protoplasts must be regarded as a *de novo* process of morphogenesis. Controlled production of spheroplasts, i.e., bodies possessing residues of incompletely digested cell walls, has been reported only in *Nadsonia elongata.*[10] However, the capacity of true protoplasts of *Nadsonia* to regenerate the cell wall was the same as in spheroplasts.[11]

2. *Biosynthesis of Components*

a. General

Yeast protoplasts have been used[12] as starting material for particulate fractions with polysaccharide synthetic activities as discussed in Volume II, Chapter 2. It has been established that in regenerating protoplasts the essential wall constituents are synthesized in quantities sufficient for the assembly of a new cell wall. However, whether these polymers differ in their chemical structures from those which make up a normal cell wall, and whether they are synthesized in the same proportions as under vegetative growth conditions are still open questions.

The lack of information is due mainly to the fact that biochemical analyses of whole populations of regenerating protoplasts is based on heterogeneous material. Freshly prepared protoplasts are largely uninuclear, but significant numbers of binuclear protoplasts and subprotoplasts without nuclei are also observed.[13,14] Protoplasts in the same population show considerable variation in their growth rate and the completion of wall regeneration is far from being synchronous. Moreover there are some protoplasts that never complete their new cell walls, and others that lyse. Nevertheless, there are some extant data on the syntheses of polysaccharide constituents and also some information on the formation of the protein moiety which have a direct bearing on cell wall regeneration.

b. Glucans

Glucans are the predominant component of microfibrils that are the first discernible structure.[15-18] Because of their physico-chemical properties microfibrils are easily isolated and purified, particularly if the protoplasts are grown in defined media that do not encourage other syntheses.[16,17] The molecular weight of the microfibrillar glucan formed by *Saccharomyces uvarum* protoplasts was estimated to be 110,000 ± 30,000.[18] Determination of the structure of glucans constituting microfibrils has been carried out on *Sacch. cerevisiae* and *Schizosaccharomyces pombe*. In both species the main component proved to be an alkali-soluble (1 → 3)-β-D-glucan.[19-21] In *Sacch. cerevisiae,* about 40% of this glucan was unbranched and of a higher degree of polymerization. The crystalline structure corresponded to paramylon rather than to hydroglucan.[20,21] The remaining glucan was probably of the same branching type as that in normal cell walls, but differed in solubility properties. In *Schiz. pombe*, in addition to (1 → 3)-β-glucan, a (1 → 3)-α-glucan has been found in X-ray diffraction patterns.[22] The synthesis of (1 → 3)-β-glucan during regeneration precedes that of α-glucan. It appears from these data that at least a part of the glucan in microfibrils in regenerated walls differs from the glucan of normal cell walls because the former is largely unbranched. Kreger and Kopecká suggest that the high proportion of linear chains is to be ascribed to partial removal of a putative β-1, 6, synthetase or a branching enzyme.[21]

The incorporation of ^{3}H-glucose into microfibrils during incubation of *Sacch. cerevisiae* protoplasts in liquid media was repressed after about 7 hr even though protein synthesis continued exponentially for a longer time.[23] On the other hand, if the assembly of microfibrils was delayed by adding snail enzymes to the medium for 24 hr, new microfibrils appeared when the enzymes were removed.[6] This suggests that the capacity of protoplasts to synthesize glucans is maintained for a very long time. In the protoplasts of *Schiz. pombe* and *N. elongata* cultivated in liquid media, the number of microfibrils increased for several days.[24,25]

2-Deoxy-D-glucose and 2-deoxy-2-fluoro-D-glucose at specific concentrations inhibit glucan synthesis.[26,27] That microfibril formation is concurrently inhibited, indicates that they are assembled from newly synthesized glucan. Papulacandin B inhibits the synthesis of the alkali-insoluble glucan fraction, but not that of alkali-soluble glucan.[28] How this inhibition is reflected in wall regeneration has not been reported. The inhibition of RNA or protein synthesis does not interfere with glucan synthesis per se.[29,30] Even if the cells are incubated with cycloheximide several hours before conversion to protoplasts, microfibril formation is not disturbed.[7] Protoplasts obtained from a number of cell cycle mutants of *Sacch. cerevisiae* form glucan fibrils at the same rate as do protoplasts of the wild type,[31] which indicates that glucan synthesis is not dependent upon progress through the cell cycle. The triggering of glucan synthesis is dependent on the supply of a metabolizable carbon source indicating either low or restricted endogenous pools.[32]

One can conclude from these data that the glucan synthetases are active in protoplasts shortly after protoplast formation, that they have a very low turnover, and that the microfibrils are formed from newly synthesized glucan. In view of these conclusions it remains to be explained why enucleated protoplasts do not form glucan microfibrils at all.[14]

c. Mannanproteins

Protoplasts retain the ability to synthesize and secrete mannans and cell wall proteins as well as the glycoprotein enzymes normally destined for the periplasmic space.[33] The latter are discussed in Volume II, Chapter 1.

Detailed analysis of mannans that compose regenerated cell walls has not yet been made, but it is of interest that freshly prepared protoplasts agglutinate in the presence of strain specific mannan antisera. This indicates[31] that the plasma membrane contains the same antigenic determinants as that of the normal cell wall (see Volume I, Chapter 6). Randomly distributed α-mannan has been detected by scanning electron microscopy on the plasma membrane of *Sacch. cerevisiae* and *Candida utilis* protoplasts using concanavalin A-colloidal gold or antimannan antibodies.[35]

The synthesis of mannan is inhibited by deoxy-glucose concentrations that are lower than those required to inhibit glucan synthesis. Consequently assembly of matrix is preferentially restrained.[36] Complete inhibition of matrix formation was found under conditions that inhibited protein synthesis but not microfibril synthesis.[30] Also, the inhibition of RNA synthesis results in cessation of matrix assembly.[29] This indicates that the messenger RNA coding for a protein or proteins involved in synthesis and/or transport of the mannanprotein moiety of the wall has a very rapid turnover.

d. Chitin

The first chemical analyses of regenerated cell walls revealed unusually high contents of glucosamine indicating the presence of chitin.[15,37] Although contamination with bud scars from the original walls casts doubt on these data, it has been confirmed that growing protoplasts synthesize *de novo* chitin. The presence of chitin in fibrillar nets produced by *Sacch. cerevisiae* protoplasts grown in liquid media has been clearly demonstrated by X-ray diffraction patterns.[20,21] The amount of chitin, estimated by specific extractions and from diffraction intensities, was up to 15% dry weight. This chitin had a very high degree of crystallinity. Farkaš and Svoboda studied the kinetics of chitin synthesis in the same system using ^{14}C-labeled glucose.[38] The initial rate of synthesis was slow, but at 3 hr the rate accelerated and by 6 hr the rate of chitin formation equalled that of glucan.

The data outlined above indicate that in protoplasts the activity of chitin synthetase is not dependent upon cytokinesis. This situation should be contrasted with the normal cell cycle in *Sacch. cerevisiae* where synthesis of chitin is discontinuous and occurs concurrently with formation of the bud scar chitin component during cytokinesis.[12]

3. Cell Wall Assembly

a. General

The first wall structure (detectable by electron microscopy) appears on the protoplast surface 20 to 60 min after the protoplasts have been transferred to nutrient medium. The reason for this delay is not known. It may be either due to a lag in the synthesis and/or transport of wall polymers or due to the fact that a certain critical concentration of polymers is required before their assembly can begin. The new structures are always microfibrillar (Figure 1). The matrix substance begins to fill in the fibrils either immediately or within several hours according to environmental conditions (Figure 2). The three-dimensional organization of the regenerating wall is determined primarily by the shape of the growing protoplast, which implies that the assembly of the wall is not under genetic control as it is in cells (see Volume II, Chapters 3 and 4).

The texture of the regenerated cell wall usually differs from the normal cell wall in several ways. The net of microfibrils is very prominent (Figure 3) so that the external surface is rarely smooth, and the new wall displays apparent layering within stained sections. In addition, remarkable aberrations in the form of local thickenings or long protuberances towards the protoplast interior are frequently observed.[39] Even under the best conditions, the time required for wall generation by protoplasts is always longer than that needed for wall replication in cells. The absolute period is much influenced by conditions of regeneration. No information exists to indicate whether protoplast-generated wall structure undergoes any further modification once reversion has started.

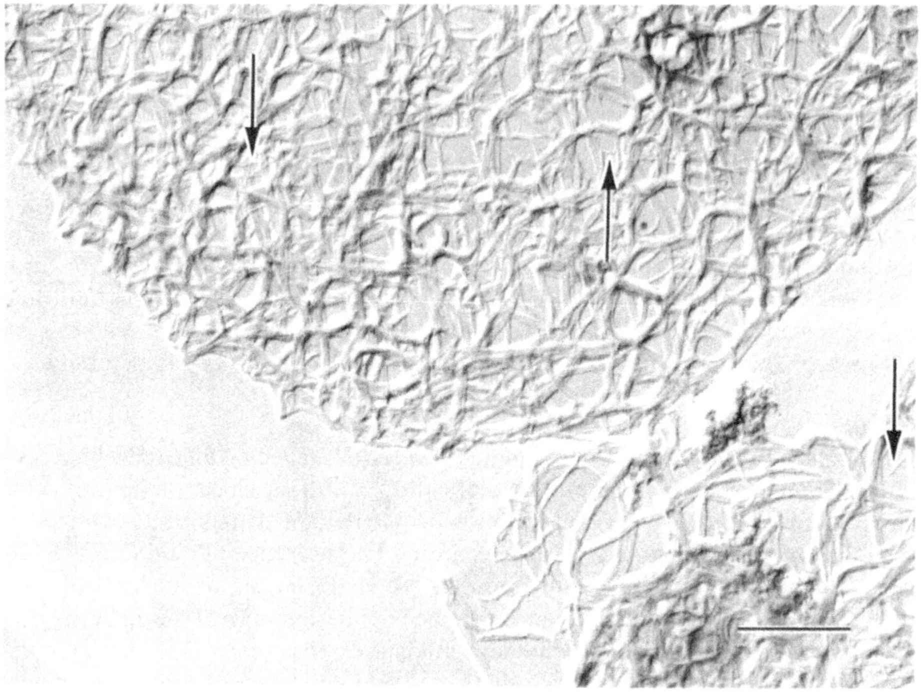

FIGURE 1. Glucan microfibrils formed by *Sacch. cerevisiae* protoplasts cultivated for 4 hr in liquid medium. Microfibrils are aggregated into broad bundles so that individual fibrils are rare (arrows). Protoplasts were lysed and the nets treated with trypsin. Electron micrograph of shadowed specimen (Courtesy of M. Kopecká). Bar equals 1 μm.

b. Regeneration Media

The first experiments[5] with yeast protoplasts showed that some factors responsible for cell wall assembly in reproducing cells are apparently absent or depleted in protoplasts. Accordingly, they must be supplied in regeneration medium; for some species they are obligatory, for others they merely accelerate wall regeneration (see Table 1).

Regeneration media for protoplasts are gels which contain 30% gelatine[5] or 2% agar.[40,41] A 30% polyethylene glycol (which is also used effectively in promoting protoplast fusion) has proved to be a successful regeneration medium.[42]

Polysaccharide microfibrils are formed in most culture media but assembly of the matrix substance occurs only in fortified media. The rate of matrix formation in gelatine media is a function of the gelatine concentration at a given temperature and of the temperature at a given gelatine concentration.[43] If the protoplasts are preincubated in a high concentration of gelatine for about 4 hr and then resuspended in liquid medium, or if (after the same incubation period) the gelatine is liquified by raising the temperature, then the assembly of the matrix still continues. These data indicate that the gelled media are mandatory only for the first steps of matrix assembly. During that time a coherent (yet thin and osmotically fragile) wall layer develops around the protoplast. In species of *Schizosaccharomyces* and *Nadsonia*, whose protoplasts can regenerate the cell wall in liquid media, the matrix is assembled after a lag of about 15 hr.[24,25,44] However, if fortified media are used the matrix begins to assemble within 30 to 60 min. We suppose that regeneration media decrease the extent of diffusion of matrix materials or precursors into the medium.[9]

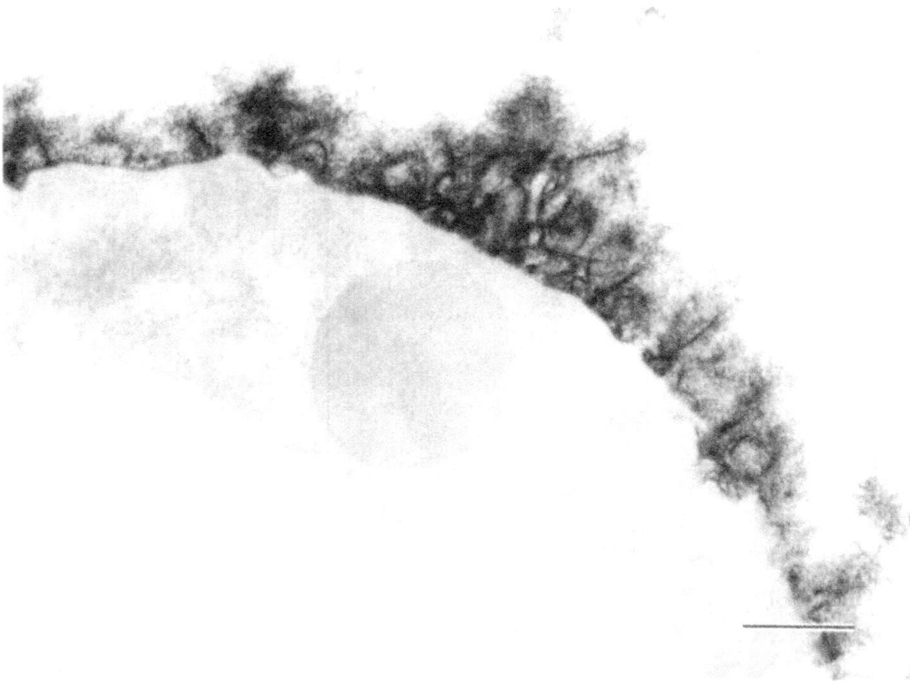

FIGURE 2. Assembly of matrix during cell wall regeneration in *Sacch. cerevisiae* protoplasts embedded in gelatine-fortified medium. Microfibrils are decorated with stained matrix substance. (Ruthenium red stain) Bar equals 0.25 μm. Transmission electron micrograph by M. Havelková.

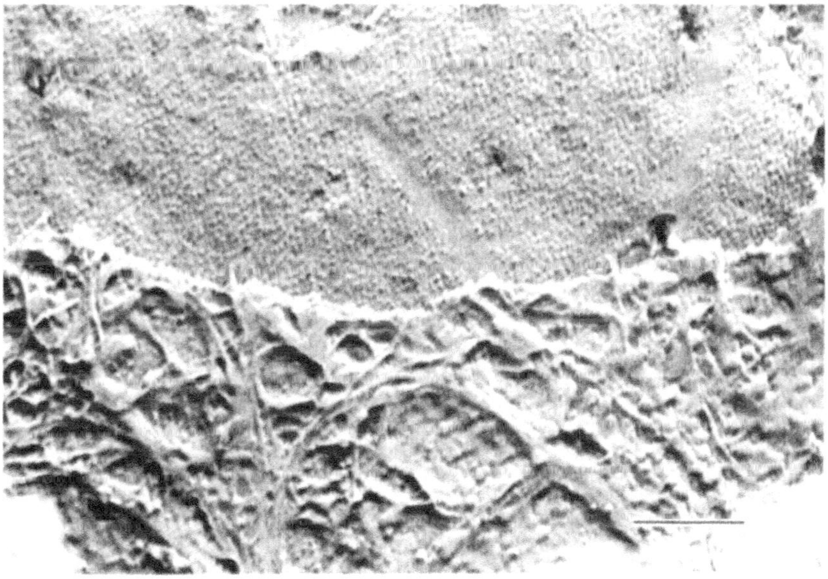

FIGURE 3. Cell wall regeneration by a *Sacch. cerevisiae* protoplast in gelatine medium. The fibrillar network is partially masked with matrix. Electron micrograph of replica of freeze-fractured and etched specimen. Bar equals 0.25 μm.

Table 1
SPECIES EXHIBITING PROTOPLAST REVERSION

Species	Special regeneration medium required?[a]	Ref.
Candida albicans	No, but accelerates reversion	71, 73
C. arborea	No	75
C. beverwijkii	No	71
C. claussenii	No	71
C. guillier mondii	No	73
C. krusei	No	71
C. tropicalis	No	73, 74
C. utilis	No, but accelerates reversion	37, 67, 69 71, 72
Endomyces magnusii	No	66
Lipomyces lipofer	No, but accelerates reversion	70
Nadsonia elongata	No, but accelerates reversion	25, 44, 67
Pichia polymorpha	No	69
Rhodotorula glutinis	Yes	67
Saccharomyces cerevisiae	Yes	See text
Sacch. chevalieri	Yes	67
Sacch. ellipsoideus	Yes	67
Sacch. fragilis	Yes	68
Sacch. lactis	Yes	67
Saccharomycodes ludwigii	Yes	67
Schizosaccharomyces pombe	No, but accelerates reversion	See text

[a] See text for details on medium.

c. Formation of Microfibrils

The diameter of a microfibril ranges from 6 to 8 nm, as measured in negatively stained preparations (Figure 4).[24,32] It is not certain whether this variation is due to the staining technique itself or whether microfibrils vary in thickness by dint of different numbers of polysaccharide chains. Even after short incubations microfibrils attain such lengths that it has not been possible to trace individual microfibrils from one end to another. The rate of fibril formation is striking. Only a few minutes after the lag period, the protoplast becomes enveloped with a net of microfibrils. The newly formed microfibrils are evenly deposited among the existing bundles over the whole protoplast surface, as shown by high resolution autoradiography.[23]

The microfibrils aggregate immediately into bundles of varying thickness (Figure 1). When measured in replicas of freeze-etched samples the spacing between microfibrils in a bundle has been found to be about 12 nm. Aggregation of microfibrils seems to be unrelated to microfibril formation in that aggregation can be influenced to a considerable degree just by the viscosity of the medium.[32]

Formation of microfibrils has also been investigated by means of freeze-fracturing which gives some insight into the structural relationship between microfibrils and the plasma membrane. This technique permits visualization of the external protoplast surface after deep etching and, depending on the fracture plane, some view of the exoplasmic and the endoplasmic fracture faces of the plasma membrane. The fracture faces reveal membrane particles which some authors claim to be functional in the

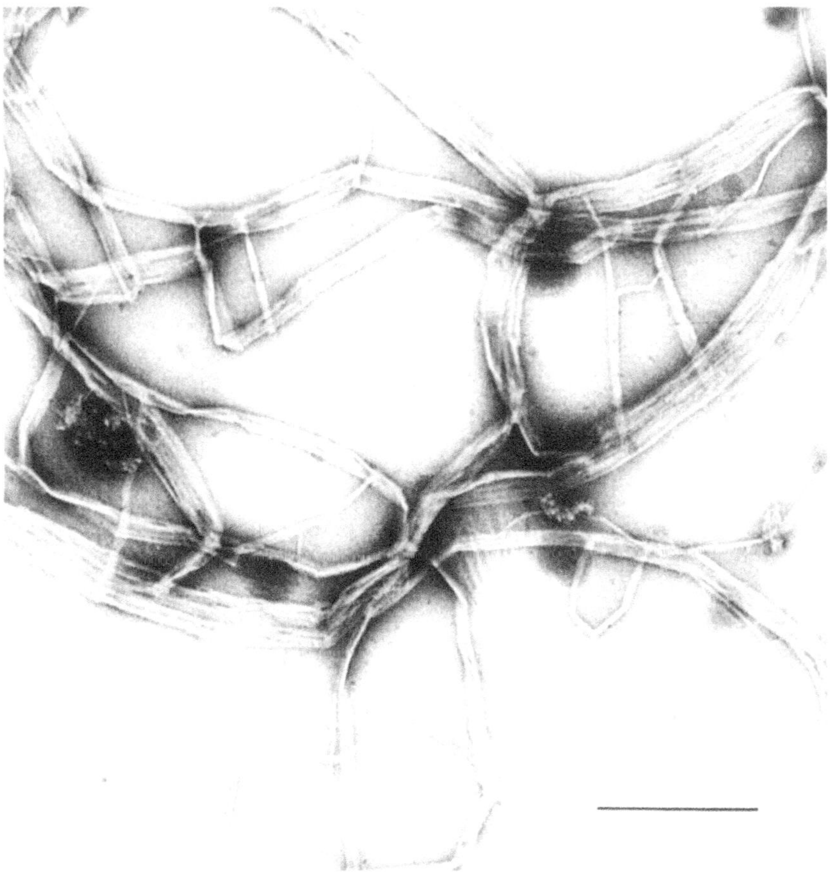

FIGURE 4. Glucan microfibrils formed by protoplasts of *Sacch. cervisiae* after negative staining with uranyl acetate. Note aggregation of microfibrils into bundles. Bar equals 0.5 μm. Electron micrograph by M. Kopecká.

formation of the skeletal component in yeast cell walls. However, the microfibrils seen in replicas after deep etching either adhere to the external surface of the plasma membrane or lie some distance from it (Figures 5 and 6).[45] If they were connected with the intramembraneous particles, they would penetrate the outer leaflet of the plasma membrane and their broken ends should be seen on the exoplasmic fracture face of the split plasma membrane. This has never been observed. The short fiber-like strucutures connecting the regenerating cell wall with the plasma membrane particles are apparently fracture artifacts.[45,46] No correlation has been found either between the orientation of microfibrils and the array of plasma membrane particles, or between the sites of their origin and hexagonally arranged particles. Similarly, the distribution of plasma membrane invaginations over the protoplast surface revealed no correlation with the origin of microfibrils.[32,45]

In sum, morphological approaches have failed to provide conclusive information that would explain the mechanism of microfibril formation in yeast protoplasts. Therefore, one can only speculate whether microfibrils arise by self-assembly from loose glucan molecules, or whether they are produced by directed (enzyme-mediated) assembly involving glucan synthetases bound to the plasma membrane.[21]

Self-assembly, both in vitro and in vivo, has been demonstrated to be a mechanism underlying the development of many cell structures. However, little is known about the

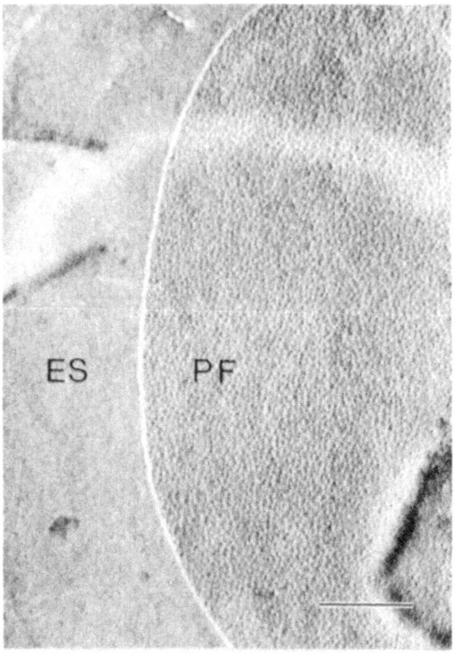

FIGURE 5. A freshly prepared protoplast of *Sacch. cerevisiae*. The external surface (ES) of the plasma membrane is free of any residues from the original cell wall. Plasma membrane particles are seen on the protoplasmic fracture face (PF). Electron micrograph of replica of freeze-fractured and etched specimen. Bar equals 0.25 μm.

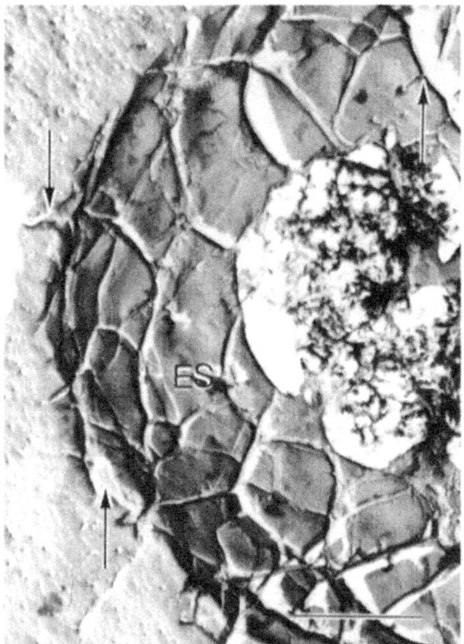

FIGURE 6. Fibrillar nets on the surface of a protoplast of *Sacch. cerevisiae*. Microfibrils lie on the external surface (ES), or are detached (arrows). Electron micrograph of replica of freeze-fractured and etched specimen. Bar equals 1 μm.

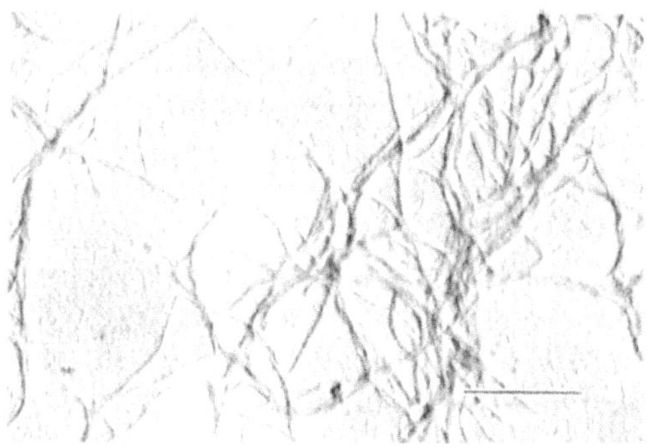

FIGURE 7. Reaggregated glucan microfibrils. Native fibrillar nets were solubilized in alkali; precipitation occurred after subsequent neutralization. Bar equals 0.5 μm. Electron micrograph of shadowed specimen by M. Kopecká.

self-assembly of structures composed of polysaccharides. Reaggregation of glucan derived from protoplast microfibrils has been demonstrated by Eddy and Woodhead,[18] and by Kopecká.[47] The fact that these reaggregated fibrils (Figure 7) differ in morphology from the native ones can be ascribed to the changed conditions in which the reaggregation occurs, and can hardly be used as an argument against the self-assembly hypothesis.

It is postulated that released glucan macromolecules, since they have a high molecular weight, accumulate around the protoplast and form a concentration gradient. As soon as the critical concentration required for crystallization is reached, microfibrils begin to assemble. The self-assembly hypothesis assumes that glucans are secreted as free molecules but makes no suggestion as to the localization of active glucan synthetases in the protoplast. These may be bound either to some intracytoplasmic structure or, as peripheral proteins, to the protoplasmic face of the plasma membrane, or they may exist as plasma membrane penetrating proteins. Substantial rates of glucan synthesis by broken plasma membranes have recently been reported (see Volume II, Chapter 2). Extracellular synthesis of glucan is not supported since microfibrils are formed in the presence of proteolytic enzymes in the medium,[48] but see Volume II, Chapter 1 for further discussion on the presence of synthetic enzymes in normal cell walls.

There are two alternatives for the directed or enzyme-mediated assembly hypothesis. First, an organized complex of glucan synthetases may be an integral part of the plasma membrane. Each enzyme of the complex would spin out one polysaccharide chain which is immediately bound to the other chains giving rise to a microfibril. This is analogous to an hypothesis suggested for the mechanism of chitin fibril formation.[49] And second, the glucan synthetases may be randomly distributed in the plasma membrane. When the synthesis of glucan starts, nascent chains would aggregate to form microfibrils. The fluidity of the plasma membrane might allow the secondary aggregation of productive enzymes to take place. In fact none of the extant hypotheses can explain fully the fact that a microfibril, which is tens of microns long, is formed within a few minutes; which would imply a polymerization rate of several hundreds of residues per second for each glucan chain. Apparently, to understand the mechanism of microfibril formation here raises similar problems to those advanced in connection with mold and higher plant protoplasts.[50] However, one advantage of the yeast protoplast system is that formation of microfibrils can be studied independently of other wall components.

d. Assembly of Matrix

Discharge of exocytotic vesicles derived from Golgi bodies is one possible mechanism for translocation of mannanproteins through the plasma membrane. They then become main constituents of wall matrix. An increase in Golgi-like bodies in *Schiz. pombe* protoplasts engaged in wall generation seems to support the view that these vesicles may be directly involved.[51,52] On the other hand, in regenerating *Sacch. cerevisiae* protoplasts no typical Golgi bodies have been demonstrated by electron microscopy, nor has the freeze-etching technique revealed any morphological signs of reverse pinocytosis in the protoplast membrane.[45] Also, the ultra-thin sections of the same protoplasts fixed *in situ* inside a gelatine medium have shown only a very small number of vesicles which could be regarded as exocytotic.[53] A remarkable finding is the hypertrophy of the endoplasmic reticulum which results in several large cisternae (at a distance of about 20 nm) from the inner side of the plasma membrane. They surround the whole protoplast periphery.[39,45,52] After cell wall regeneration has finished and reversion has started, the cisternae in question retract from the periphery. This behavior of the endoplasmic reticulum may be related to the synthesis or translocation of wall matrix material. It should be noted that the majority of mannan synthesizing activity has been found in the fraction of vesicles derived from the endoplasmic reticulum.[12,54]

In protoplasts which can regenerate the cell wall in liquid media, only fibrils are organized during the first 15 hr, and they form a very broad and dense basket around the protoplast.[24,25,52,55] Subsequently the matrix substance appears near the plasma membrane (Figure 9). Evidently, the dense layer of microfibrils prevents the loss of matrix components into the medium and thus functions in a similar way to the artificial matrix of gelled medium.

If the protoplasts of regenerating *Sacch. cerevisiae* are closely packed in gelatine, common cell walls are organized around the contiguous protoplasts.[39,56] In that case the matrix material is apparently assembled as one structural unit regardless of origin. Thus, matrix assembly seems to be a self-assembly process which results from the physico-chemical properties of mannanprotein complexes rather than a process which is under direct cell control.

e. Inhibition of Wall Assembly

Cell wall regeneration can be inhibited at several levels. The inhibition of synthesis of wall polymers has already been mentioned. Inhibition at the level of translocation of wall polymers is poorly documented. Cytochalasin B (an inhibitor of exocytosis) has no effect on wall regeneration.[57] Inhibition at the level of assembly of wall structures is discussed below.

Wall formation in *Sacch. cerevisiae* is abolished by adding digestive enzymes of the snail to culture medium.[6] This indicates that secreted wall components or at least key components are immediately digested by the snail enzymes. In this case the protoplasts grow into giant, multinuclear forms. If the protoplasts are now transferred to an enzyme-free medium, wall regeneration starts and culminates in the formation of fibrillar nets in liquid media, and of complete cell walls in gelled media. The blockage of cell wall regeneration by lytic enzymes is reversible at any time. The same is not true for protoplast reversion. If the blockage of wall regeneration lasts longer than about 6 hr, protoplasts regenerate their cell walls but inexplicably never revert to cells and they die within several days.

Different results were obtained with *Schiz. pombe* and *N. elongata*.[58] Here, blockage of wall formation operates only during the first 15 to 24 hr. Thereafter, aberrant wall structures are obtained. In *Schiz. pombe*, they are composed of short microfibrils masked with some matrix substance and in *N. elongata* they consist of long microfibrils.

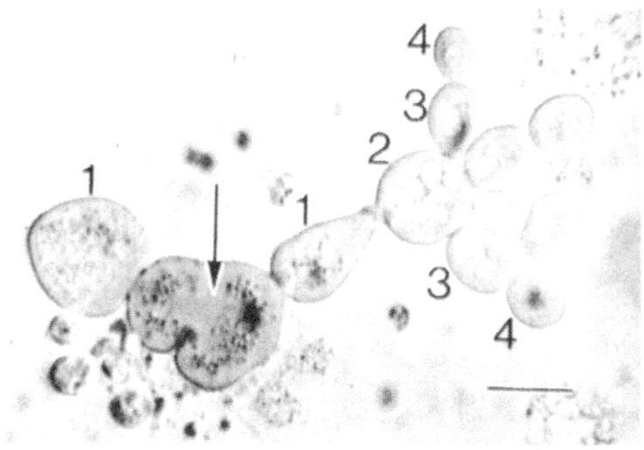

FIGURE 8. Reversion of a *Sacch. cerevisiae* protoplast. The original protoplast (now with new wall) is marked by arrow. Four generations of revertants (numbered sequentially 1 to 4) are indicated. Phase-contrast micrograph. Bar equals 10 μm.

Apparently, these structures are resistant to the snail gut enzymes. Some *Schizosaccharomyces* protoplasts with the aberrant walls may even undergo septation, but in the presence of lytic enzymes true reversion never occurs. If, however, the blockage is removed by transferring the protoplasts to an enzyme-free medium, wall regeneration is completed and reversion begins. Inhibition of reversion in these two species is reversible at any time which contrasts with the situation in *Sacch. cerevisiae.*

C. Protoplast Reversion

As soon as the regenerated cell wall achieves sufficient complexity (the degree of which is still unknown) processes are initiated which lead to reversion, i.e., to the step-by-step recovery of the normal cell cycle. The morphology of reversion depends on several factors, the most important of which are the species-specific mechanism of cytokinesis, and the history of the protoplast, i.e., the number of nuclear cycles completed during the growth period.

1. Morphology of Reversion

a. Saccharomyces cerevisiae

The morphology of reversion in protoplasts of *Sacch. cerevisiae* (following wall regeneration) corresponds in principle to the process of budding.[39] The budding is often multipolar, especially in protoplasts containing several nuclei. Cells of the first generation of revertants are usually bigger than normal cells and are frequently of irregular shape (Figure 8). Often more than one nucleus passes into the bud from the mother "protoplast". The septum usually has structural aberrations in the form of plugs protruding on both sides. These abnormalities indicate that although the basic cytokinetic mechanism is functioning, its individual steps are not properly coordinated.

The next generation of revertants has a morphology almost identical to that of the original strain. However, in some cases the species-specific shape may not be achieved until later generations. Rarely, a protoplast produces a stable revertant with different cell shape or doubling time.[59] Some multinucleate revertants were able to maintain this state of affairs for thousands of generations.[60,61] These unusual revertants require more detailed studies in order to decide whether they are mutants or modulations controlled by some epigenic system.

b. Schizosaccharomyces pombe

Schiz. pombe protoplasts are spherical and reversion also starts from spherical protoplasts with fresh walls.[24] The first cell division occurs by formation of a transverse

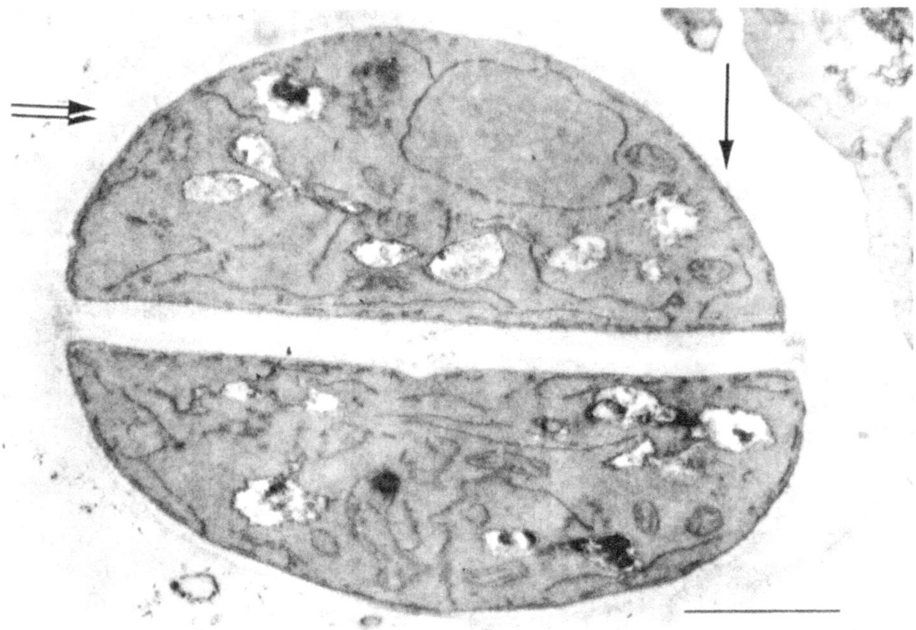

FIGURE 9. First cell division of reverting *Schiz. pombe* protoplast. The septum divides the spherical, newly walled protoplast into symmetric parts. The regenerated wall (arrow) is surrounded by a broad zone of microfibrils (double arrows). Bar equals 1 μm. Transmission electron micrograph by M. Havelková.

septum growing from the periphery and dividing the walled protoplast usually into two symmetric half-spheres (Figure 9). The subsequent generation of revertants, however, is not produced by fission (see Volume II, Chapter 4) but by a process resembling budding. Apparently, the regenerated wall possesses a site more analogous to the primary pole of the normal cell wall.

If the growth period is prolonged by blocking cell wall regeneration, multinuclear bodies similar to those sometimes encountered in *Sacch. cerevisiae* result.[58,62] During reversion each multinuclear body produces several septa that are irregular both in morphology and in location (Figure 10). No convincing evidence has yet been provided to correlate number of septa with total number of nuclei or even with the number of nuclei undergoing division. Likewise the positions of septa may be guided by the orientation of nuclear plaques but the data are scant.

c. *Dimorphic Yeasts*

The dimorphic yeasts exist in different cell forms according to growth conditions. *Saccharomycopsis fibuligera* cells grow in mycelial and yeast forms, *Trigonopsis variabilis* in pyramidal and ellipsoidal forms. Their protoplasts, regardless of the shape of parental cells, are spherical and remain so during cell wall regeneration, apparently because environmental factors controlling the dimorphic morphology in normal cells are not involved.[63,64] These factors, however, exert their effect on reversion, i.e., on the morphology of the first generation of revertants. According to culture conditions, *S. fibuligera* protoplasts revert to mycelial or yeast forms, and protoplasts of *Trig. variabilis* (after some transient stage) revert to triangular or ellipsoidal forms. These results suggest either that protoplasts cannot respond to specific signals determining dimorphic character or, more likely, that the morphology of the regenerating wall cannot accommodate these peculiarities whereas walls of first or second generation revertants can.

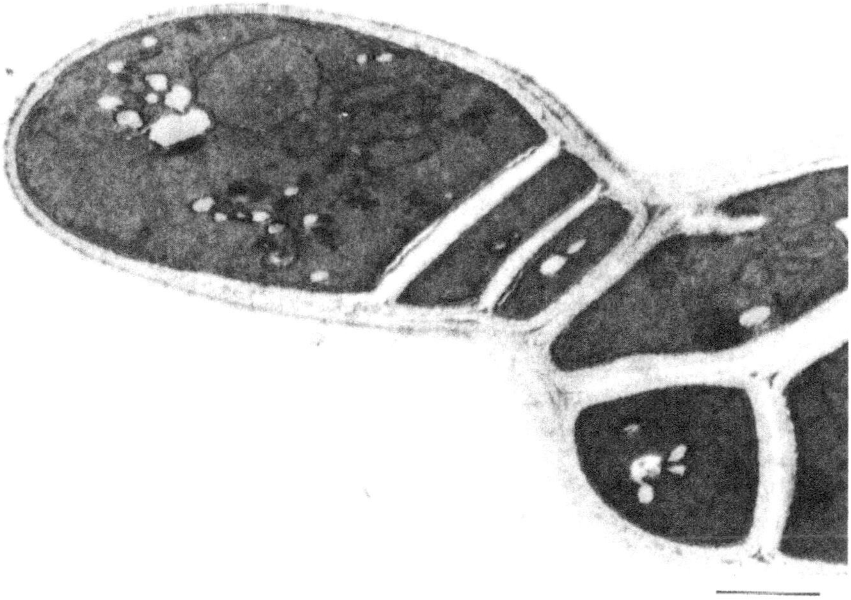

FIGURE 10. Reversion of a *Schiz. pombe* protoplast containing several nuclei (not visible in this plane) and multiple septations. Protoplasts were cultivated 24 hr in the presence of snail enzymes and then transferred to enzyme-free medium where cell wall regeneration and reversion took place. Bar equals 1 μm. Transmission electron micrograph by M. Havelková.

d. Other Yeasts

Wall generation by protoplasts and reversion have been described in a number of yeasts belonging to different taxonomic groups. All these observations are in general agreement with the patterns described for *Sacch. cerevisiae* or *Schiz. pombe.*[9,65] A compilation of tested species is assembled in Table 1.

2. Morphogenetic Role of the Cell Wall

The fibrillar network formed in liquid media or after selective inhibition of matrix assembly has no morphogenetic function.[9] It is not able to induce cytokinesis unless complemented with the matrix substance. On the other hand, the fibrillar meshwork on the protoplast surface may be involved in determining the shape of a growing protoplast in another fashion. The increasing volume of the cytoplasm may cause a break in the fibrillar network, and thus produce a naked hernia that imitates polar growth or budding (Figure 11).[16,25,32] Sometimes the hernia, which may contain a nucleus, splits off, i.e., reminiscent of division. When protoplasts grow on the surface of solid medium, the frequency of this "pseudodivision" is increased.[76] Evidently, other physical factors, such as surface tension or adhesion to the supporting gel, are involved. All these phenomena of protoplast "budding" or "pseudodivision" can be mistaken for cell controlled processes, though in fact they are only physical fragmentations of growing protoplasts.

The regenerated cell wall follows the shape of the protoplast. Apparently, the genetic signals controlling the species-specific wall morphology are out of operation or cannot be recognized by the mechanisms of wall repair. The same is true for regenerating protoplasts of dimorphic yeasts. In this sense, the regenerating cell wall behaves as a morphogenetically passive structure. The genetically controlled assembly of the cell wall comes into operation only in reverting cells. However, it is not known whether the regenerated wall is a prerequisite for this function as a kind of epigenic template or

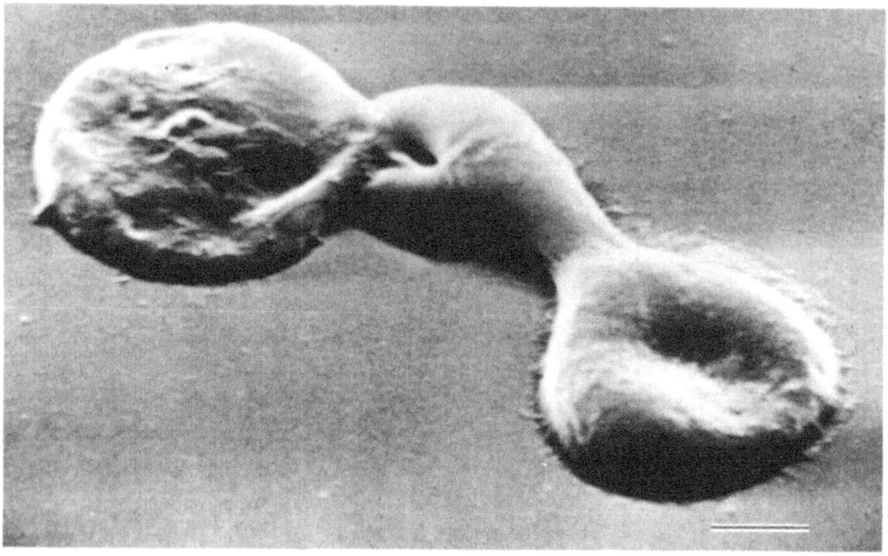

FIGURE 11. A *Sacch. cerevisiae* protoplast grown in liquid medium. Herniation probably resulted from local weakness in the fibrillar network. Scanning electron micrograph. Bar equals 2 μm.

whether genetically controlled assembly of the cell wall is made possible by a repair in coordination between cell cycle events during reversion. The cell wall in revertants assumes an active morphogenetic function in cell shape determination, although this may not be fully expressed until the third or fourth generation.

The activity of the cytoskeleton is thought to play a role in cell shape determination. The facts outlined above may indicate that microtubules and microfilaments are able to operate only in close coordination with a complex cell wall. In any case, the system of reverting protoplasts seems to be a promising tool in identification of signals for cell shape.

Protoplasts derived from diploid yeasts may sporulate under conditions in which cell wall regeneration is prevented.[15,77] This implies that both meiotic division and spore wall formation are processes independent of the presence of a regenerated ascus wall. The initiation of mating in haploids of opposite mating-type depends on the integrity of the cell wall.[78] Sex-specific morphogenetic changes do not occur in protoplasts but may happen after the regeneration of the cell wall. However, protoplasts are capable of inducing sexual activation in normal cells of opposite mating-type.

Yeast protoplasts cultivated in a liquid media unsupportive of wall regeneration soon undergo lysis, which may be of metabolic nature.[79] Protoplasts maintained on the surface of solid media may not regenerate walls but they tend to live longer, for many days and even weeks. When they eventually break they form fragments of different sizes with varying numbers of nuclei.[76] These fragments also have good longevity. To date all attempts at unlimited propagation of protoplasts (without wall regeneration) have failed, as has been the case for the protoplasts of other eukaryotic cells and most bacteria. The only exceptions to this rule are stable bacterial L-forms. It will be desirable to continue experimentation along these lines since it may provide information essential to the possible teleonomic role of the cell wall.

II. PROTOPLAST FUSION

A. Introduction

The knowledge that has accumulated to date on the morphogenesis of protoplasts has provided a basis for further investigation into the genetic and epigenetic control of morphogenetic processes both at the cellular and the protoplast level. One of the possible experimental approaches to this problem is genome manipulation at the cellular level, i.e., through fusion.

For such studies, yeast protoplasts appear to be very suitable models since they are easy to isolate and propagate. Moreover, a large number of mutants exists and this facilitates the isolation of the fusion products.

In the late 1960s spontaneous fusion of yeast protoplasts was noticed occasionally, but results were not reproducible.[80] Fusion of yeast protoplasts became a practical technique only after the introduction of polyethylene glycol (PEG) which had previously proved to be a powerful fusogen for protoplasts of higher plants.[81] The first successful fusion reported by Sipiczki and Ferenczy for *Schiz. pombe*[82] was soon followed by a series of reports on *Sacch. cerevisiae,*[83–87] *Kluyveromyces lactis,*[88] *Saccharomycopsis lipolytica,*[89] *C. tropicalis,*[90] and *Rhodosporidium toruloides.*[91] Intergeneric fusion between *Saccharomycopsis fibuligera* and *C. tropicalis*[92] and even an interkingdom hybridization between yeast protoplasts and hen erythrocytes have been reported.[93] These examples show that the fusion of protoplasts is not only possible but can be experimentally controlled. In contrast to protoplasts of higher plants, fusion frequencies in yeasts are considerably lower, varying between 0.00001 and 10%. Improvements in technique are expected to improve this. However, this disadvantage is outweighed by the wide choice of stable genetically marked strains which enable the analyses to be carried out with reliable results.

B. Fusion Procedure for Hybrids

1. Technique

In order to ensure selective proliferation of fused protoplasts auxotrophic strains are usually chosen as source organisms. Through complementation of their genotypes they can give rise to a fusion product which, in contrast to the parental protoplasts, can grow on minimal media. The restoration of prototrophy, i.e., the expression of biochemical markers, is evident already during cell wall regeneration and reversion, whereas the expression of other markers such as mating-type or morphological markers can be demonstrated only in a population of reverted hybrid cells. Because of a relatively low yield of reversion of protoplasts to cells, it is advisable to employ parents with one, or no more than two, auxotrophic markers so that the vitality of the hybrid protoplasts will not be reduced.

For fusion, parental protoplasts (about 10^6 to 10^7 of each) are mixed and pelleted. The sediment is suspended in 25 to 40% (w/v) PEG (mol wt 4000 or 6000) supplemented with 0.01 to 0.1 *M* $CaCl_2$. The length of incubation with PEG (20 to 30 min), PEG molarity, and incubation temperature (20 to 37° C) have no significant effect on the yield of fused pairs. The suspension is then diluted with osmotically stabilized regeneration medium containing 2 to 3% agar.[40] Agar medium has proved to be the best environment for promotion of wall regeneration as well as for selection of hybrids. Prototrophic colonies can be detected under the light microscope after 1 day; visible colonies appear at 2 to 3 days of cultivation in the regeneration medium. The frequency of fusion is expressed as a ratio of the number of prototrophic colonies on minimal medium to that of colonies arising from reverted protoplasts on complete medium.

2. Mechanism

Addition of PEG to pelleted protoplasts results in their strong agglutination. The close contact between protoplasts is maintained throughout the incubation with PEG although the aggregates can be disrupted at any time by mere dilution of the fusogen. A slow dilution of the aggregated protoplasts with osmotically stabilized regeneration medium results in the appearance of multinuclear bodies, suggesting that fusion has taken place. The time course of the fusion process has been studied by electron microscopy.[94] First, a close apposition of partner plasma membranes occurs immediately after PEG is added (Figure 12). Incubation of protoplasts with the fusogen leads to ultrastructural changes in the membranes at the sites of their contact; the membranes coalesce forming a common penta- or trilaminar membrane characterized by a considerable reduction in the number of intramembraneous particles. Then, during incubation in regeneration medium, these areas of adhered membranes show protrusions with crater-like openings, which are considered to be initial channels connecting adjoining protoplasts. These pores may later enlarge to form one broad connection. Apparently the function of the PEG is to induce contact on a broad base, but it does not seem to influence the secondary events. Only after the membranes have been in close apposition for a prolonged period, do the changes in ultrastructure described above take place. The cytoplasmic fusion between the partners requires the presence of the nutrient regeneration medium, suggesting that the protoplasts should be in an active metabolic state.[94]

3. Characterization of Hybrids

Hybrid protoplasts are capable of proliferating in minimal media. Depending on culture conditions and the species involved they either grow into large vacuolated bodies (Figure 13) or regenerate the walls and revert to cells.[84] It has been found that the frequency of growing protoplasts is at least one order higher than the frequency of fusions estimated from reversion. This indicates that not all fused protoplasts produce prototrophic revertants or, in other words, that the fusion of complementary genomes does not ensure protoplast reversion.[84]

The morphogenesis of hybrid protoplasts during growth and reversion does not differ appreciably from the morphogenesis of the relevant parental protoplasts, although this has only been studied so far in intraspecific fusion experiments.[84] More information on cell wall regeneration and reversion may emerge from interspecific hybridization experiments, particularly in such instances where parental protoplasts differ in their requirements for regeneration media.

The size of hybrid cells produced by intraspecific fusion is generally larger than the size of either of the parental organisms; occasionally, it may even exceed the sum of the volumes of the parents.[83-87] The DNA content of the hybrid cell varies between diploidy and tetraploidy, but values lower than those corresponding to diploids have also been reported.[83,85,91] Each hybrid cell contains only one nucleus, which suggests that fusion of cytoplasms was followed by karyogamy.[82,85,87] Of interest are the results of interspecific hybridization between *Schiz. pombe* and *Schiz. versatilis.*[95] The hybrids are spherical and osmotically sensitive. Larger hybrids (in comparison with the parental strains) have been reported for other interspecific and intergeneric combinations.[96,97] The hybrid cells of *S. fibuligera* and *C. tropicalis* did not show novel morphology.[92] In sum, the number of cases is too small to make a categorical statement.

Fusion of opposite mating-type protoplasts results in formation of diploid cells. These have the same sporulation capacity as cells produced by conjugation of haploid cells.[84,85] There is no difference in fusion frequencies between protoplasts of the opposite mating-type and those of the identical mating-type.[85,89] This finding implies that the

FIGURE 12. Agglutination of *Sacch. cerevisiae* protoplasts induced by 35% polyethylene glycol. The plasma membranes are closely apposed and no breaks in the continuity are visible. Electron micrograph of a replica of a freeze-fractured and etched specimen. Bar equals 2 μm.

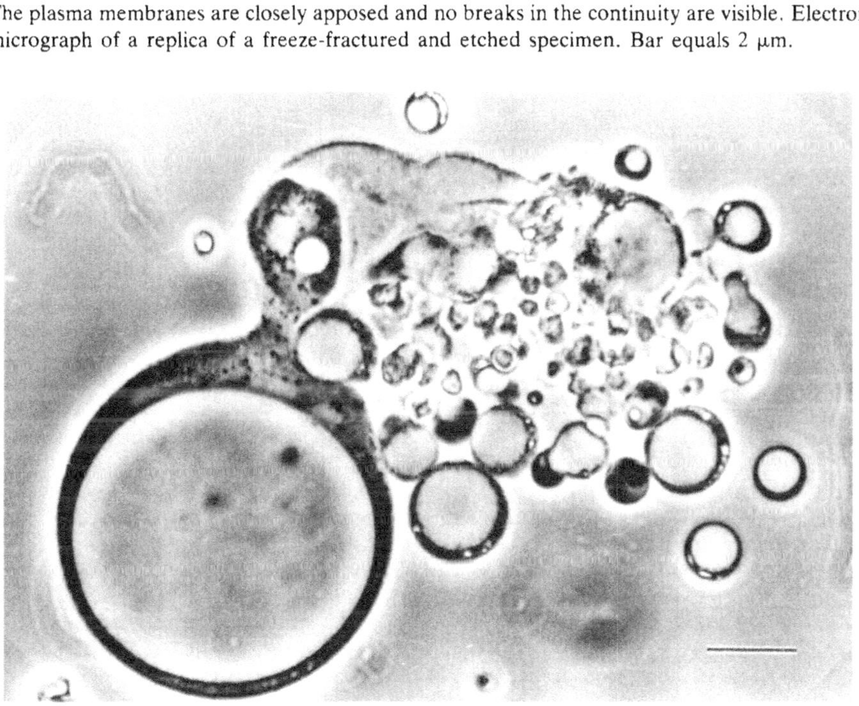

FIGURE 13. Hybrid protoplast and several resting protoplasts. A mixture of complementary auxotrophs of *Sacch. cerevisiae* was treated with polyethylene glycol; the fusion product was able to grow on minimal medium. Phase-contrast micrograph. Bar equals 10 μm.

mating characters have no effect on the potential of protoplasts to fuse. The genome of hybrid cells may include genetic markers of two or even three parental cells, as demonstrated by mitotic and meiotic segregation analyses.[82-87] The amount of genetic material in hybrids varies between diploidy and tetraploidy but various stages of aneuploidy have also been observed.[91] In addition to the transfer of nuclear genes via fusion, the transfer of mitochondrial genes has also been reported.[83,88]

C. Prospects

Extant results on hybridization by protoplast fusion indicate that this technique has a promising future. As in animal and higher plant cells, this method offers new possibilities for genetics. Moreover, fusion opens the door for the introduction of imperfect yeasts or polyploid industrial strains into genetic research. For morphogenetic studies, the fusion technique is a promising tool for investigating interactions between different kinds of morphogenetic markers, different modes of growth (polarized vs. nonpolarized), and different types of cell division (budding vs. fission). It may be anticipated that the construction of hybrid genomes will be useful in the study of cell surface characteristics, such as flocculation and fragility, which are important in industrial applications. The new combinations might also show some favorable biochemical characteristics, such as an increased utilization of substrates, production of utile primary and secondary metabolites, and an enhanced formation of biomass. However, all this will first require an increase in fusion efficiency, and a better understanding of both fusion mechanisms and the processes of integration and segregation of the parental genomes.

ACKNOWLEDGMENTS

We thank Z. Doležalová, E. Streiblová, J. Chaloupka, and P. Nurse for comments on this chapter. M. Havelková and M. Kopecká kindly provided micrographs.

REFERENCES

1. **Nečas, O.,** Regeneration of protoplasts, in *Advances in Protoplast Research*, Farkaš, G. L. and Ferenczy, L. Eds., Akademiai Kiado, Budapest, 1980, 151.
2. **Nečas, O.,** The vitality of cell fragments of yeasts. I. Plasmatic droplets of yeasts, *Folia Biol. (Praha)*, 1, 19, 1955.
3. **Nečas, O,** The vitality of cell fragments of yeasts. III. Regeneration of cells from plasmatic formations, *Folia Biol. (Praha)*, 1, 220, 1955.
4. **Nečas, O.,** Regeneration of yeast cells from naked protoplasts, *Nature (London)*, 177, 898, 1956.
5. **Nečas, O.,** Physical conditions as important factors for the regeneration of naked yeast protoplasts, *Nature (London)*, 192, 580, 1961.
6. **Svoboda, A. and Nečas, O.,** Mechanism of regeneration of yeast protoplasts. VI. An experimental blocking of regeneration of protoplasts, *Folia Biol. (Praha)*, 14, 390, 1968.
7. **Svoboda, A. and Nečas, O.,** Experimental decoupling in the synthesis of fibrillar and amorphous components during regeneration of the cell wall in *Saccharomyces cerevisiae* protoplasts, in *Yeast Protoplasts*, Nečas, O. and Svoboda, A., Eds., J. E. Purkyně University, Brno, 1970, 211.
8. **Nečas, O., Janisch, R., Jahoda, J., and Gabriel, M.,** Division of the nuclei of naked yeast protoplasts, *Folia Biol. (Praha)*, 7, 202, 1961.
9. **Nečas, O.,** Cell wall synthesis in yeast protoplasts, *Bact. Rev.*, 35, 149, 1971.
10. **Havelková, M.,** Controlled production of sphaeroplasts in the yeast *Nadsonia elongata, Arch. Microbiol.*, 90, 77, 1973.

11. **Havelková, M.**, Reversion of protoplasts and sphaeroplasts in the yeast *Nadsonia elongata, Arch. Microbiol.*, 93, 165, 1973.
12. **Farkaš, V.**, Biosynthesis of cell walls of fungi, *Microbiol. Microbiol. Rev.*, 43, 117, 1979.
13. **Ottolenghi, P.**, The numerical equivalence of cells and protoplasts in a budding yeast, *Exp. Cell Res.*, 43, 148, 1966.
14. **Kopecká, M., Gabriel, M., and Nečas, O.**, A method of isolating unnucleated yeast protoplasts unable to synthesize the glucan fibrillar component of the wall, *J. Gen. Microbiol.*, 81, 111, 1974.
15. **Eddy, A. A. and Williamson, D. H.**, Formation of aberrant cell walls and of spores by growing yeast protoplasts, *Nature (London)*, 183, 1252, 1959.
16. **Nečas, O.**, The mechanism of regeneration of yeast protoplasts. II. Formation of the cell wall *de novo, Folia Biol. (Praha)*, 11, 97, 1965.
17. **Kopecká, M., Čtvrtniček, O., and Nečas, O.**, Formation and properties of the fibrillar network formed in yeast protoplasts as the first step of biosynthesis of cell wall, in *Symposium über Hefeprotoplasten*, Müller, R., Ed., Akademie-Verlag, Berlin, 1967, 73.
18. **Eddy, A. A. and Woodhead, J. S.**, An alkali-soluble glucan fraction from the cell walls of the yeast *Saccharomyces carlsbergensis, FEBS Lett.*, 1, 67, 1968.
19. **Kreger, D. R. and Kopecká, M.**, On the nature of the fibrillar nets formed by protoplasts of *Saccharomyces cerevisiae* in liquid media, in *Yeast, Mold and Plant Protoplasts*, Villanueva, J. R. Garcia-Acha, I., Gascón, S., and Uruburu, F., Eds., Academic Press, London, 1973, 117.
20. **Kreger, D. R. and Kopecká, M.**, On the nature and formation of the fibrillar nets produced by protoplasts of *Saccharomyces cerevisiae* in liquid media. An electronmicroscopic, X-ray diffraction and chemical study, *J. Gen. Microbiol.*, 92, 207, 1976.
21. **Kreger, D. R. and Kopecká, M.**, Assembly of wall polymers during the regeneration of yeast protoplasts, in *Microbial and Plant Protoplasts*, Peberdy, J., Rose, A. H., Rogers, H. J., and Cocking, E. C., Eds., Academic Press, London, 1976, 237.
22. **Kreger, D. R. and Kopecká, M.**, The nature of nets produced by protoplasts of *Schizosaccharomyces pombe* during the first stage of wall regeneration in liquid media, *J. Gen. Microbiol.*, 108, 269, 1978.
23. **Nečas, O., Soška, J., Kopecká, M. and Reich, J.**, Application of ^{3}H-glucose in the study of cell wall synthesis in yeast protoplasts, in *Yeast Protoplasts*, Nečas, O. and Svoboda, A., Eds, J. E. Purkyně University, Brno, 1970, 203.
24. **Nečas, O., Svoboda, A., and Havelková, M.**, Mechanism of regeneration of yeast protoplasts. V. Formation of the cell wall in *Schizosaccharomyces pombe, Folia Biol. (Praha)*, 14, 80, 1968.
25. **Havelková, M.**, Mechanism of regeneration of yeast protoplasts. VII. Electron microscopy of growing and regenerating protoplasts of *Nadsonia elongata, Folia Microbiol. (Praha)*, 14, 462, 1969.
26. **Farkaš, V., Svoboda, A., and Bauer, Š.**, Inhibitory effect of 2-deoxy-D-glucose on the formation of the cell wall in yeast protoplasts, *J. Bacteriol.*, 98, 744, 1969.
27. **Biely, P., Kovařík, J., and Bauer, Š.**, Lysis of *Saccharomyces cerevisiae* with 2-deoxy-2-fluoro-D-glucose, an inhibitor of the cell wall glucan synthesis, *J. Bacteriol.*, 115, 1108, 1973.
28. **Baguley, B. C., Römmele, G., Gruner, J., and Wehrli, W.**, Papulacandin B: an inhibitor of glucan synthesis in yeast sphaeroplasts, *J. Biochem.*, 97, 345, 1979.
29. **Kopecká, M. and Farkaš, V.**, RNA synthesis and the formation of cell wall. Effect of lomofungin on regenerating protoplasts of *Saccharomyces cerevisiae, J. Gen. Microbiol.*, 110, 453, 1979.
30. **Nečas, O., Svoboda, A. and Kopecká, M.**, The effect of cycloheximid (actidion) on cell wall synthesis in yeast protoplasts, *Exp. Cell Res.*, 53, 291, 1968.
31. **Kopecká, M.**, Biogenesis of the fibrillar wall component in the cell cycle of *Saccharomyces cerevisiae, Folia Microbiol. (Praha)*, 22, 440, 1977.
32. **Kopecká, M., Nečas, O., and Svoboda, A.**, Structure and formation of the fibrillar component of the growing wall in yeast protoplasts, in *Yeast Protoplasts*, Nečas, O. and Svoboda, A., Eds., J. E. Purkyně University, Brno, 1970, 197.
33. **Lampen, J. O., Kuo, S. C., and Cano, F. R.**, Control of synthesis and secretion of exoenzymes by yeast protoplasts, in *Yeast, Mold and Plant Protoplasts*, Villanueva, J. R., Garcia-Acha, I., Gascón, S., and Uruburu, F., Eds. Academic Press, London 1973, 143.
34. **Ballou, C. and Svoboda, A.**, personal communication.
35. **Horisberger, M., Rosset, J., and Bauer, H.**, Localization of mannan at the surface of yeast protoplasts by scanning electron microscopy, *Arch. Microbiol.*, 109, 9, 1976.
36. **Farkaš, V., Svoboda, A. and Bauer, Š.**, Secretion of cell-wall glycoproteins by yeast protoplasts. Effect of 2-deoxy-D-glucose and cycloheximide, *Biochem. J.*, 118, 755, 1970.
37. **Novaes-Lediau, M. and Garcia-Mendosa, G.**, Biochemical studies on wall synthesized by *Candida utilis, J. Gen. Microbiol.*, 61, 335, 1970.
38. **Farkaš, V. and Svoboda, A.**, Kinetics of β-glucan and chitin formation by cells and protoplasts of the yeast *Saccharomyces cerevisiae, Curr. Microbiol.*, 4, 99, 1980.

39. **Nečas, O. and Svoboda, A.,** Mechanism of regeneration of yeast protoplasts. IV. Electron microscopy of regenerating protoplasts, *Folia Biol. (Praha),* 13, 379, 1967.
40. **Svoboda, A.,** Regeneration of yeast protoplasts in agar gels, *Exp. Cell Res.,* 44, 640, 1967.
41. **Rost, K.,** Gewinnung und Eigenschaften von Protoplasten aus *Schizosaccharomyces pombe, Z. Allg. Mikrobiol.,* 9, 289, 1969.
42. **Svoboda, A. and Piedra, D.,** unpublished data.
43. **Nečas, O. and Svoboda, A.,** Relation between the biosynthesis of the cell wall and regeneration in yeast protoplasts, in *Symposium über Hefeprotoplasten,* Müller, R., Ed., Akademie-Verlag, Berlin 1967, 67.
44. **Havelková, M. and Svoboda, A.,** Regeneration of *Nadsonia elongata* protoplasts, in *Yeast Protoplasts,* Nečas, O. and Svoboda, A., Eds., J. E. Purkyně University, Brno, 1970, 43.
45. **Nečas, O. and Svoboda, A.,** Regeneration of yeast protoplasts. A freeze-etching study, *Z. Allg. Mikrobiol.,* 16, 615, 1976.
46. **Sleytr, U.,** Interpretation of morphological data obtained by freeze-fracture replication methods, *Acta Histochem., Suppl.,* 23, 121, 1981.
47. **Kopecká, M.,** Reaggregation of a fibrillar component of the cell wall of yeast protoplasts, *Folia Microbiol. (Praha),* 18, 159, 1975.
48. **Nečas, O. and Svoboda, A.,** Effect of proteases, phospholipases and polysaccharide-splitting enzymes on plasma membrane particles and on the synthesis of the fibrillar cell wall component in yeast protoplasts, *Folia Microbiol. (Praha),* 19, 81, 1974.
49. **Bartnicki-Garcia, S., Bracker, C. E., Reyes, E., and Ruiz-Herrera, J.,** Isolation of chitosomes from taxonomically diverse fungi and synthesis of chitin microfibrils in vitro, *Exp. Mycol.,* 2, 173, 1978.
50. **Willison, J. H. M.,** Synthesis of cell wall by higher plant protoplasts, in *Microbial and Plant Protoplasts,* Peberdy, J. F., Rose, A. H., Rogers, H. J., and Cocking, E. C., Eds., Academic Press, London, 1976, 283.
51. **Havelková, M. and Menšík, P.,** The Golgi apparatus in the regenerating protoplasts of *Schizosaccharomyces, Naturwissenschaften,* 53, 562, 1966.
52. **Havelková M.,** Electron microscopy study of cell structures and their changes during growth and regeneration of *Schizosaccharomyces pombe* protoplasts, *Folia Microbiol. (Praha),* 14, 155, 1969.
53. **Havelková, M.,** personal communication.
54. **Ballou, C.,** Structure and biosynthesis of the mannan component of the yeast cell envelope, *Adv. Microb. Physiol.,* 14, 93, 1976.
55. **Svoboda, A., Nečas, O., and Havelková, M.,** Biosynthesis of cell wall in protoplasts of *Schizosaccharomyces pombe,* in *Proc. 2nd Symp. on Yeast,* Kocková-Kratochvílová, A., Ed., Slovak Academy of Science, Bratislava, 1969, 219.
56. **Nečas, O. and Svoboda, A.,** Electron microscopy of regenerating yeast protoplasts, in *Proc. 2nd Symp. on Yeast,* Kocková-Kratochvílová, Ed., Slovak Academy of Science, Bratislava, 1969, 213.
57. **Nečas, O. and Svoboda, A.,** unpublished data.
58. **Havelková, M.,** Experimental inhibition of cell wall formation and of reversion in *Nadsonia elongata* and *Schizosaccharomyces pombe* protoplasts, *Protoplasma,* 75, 405, 1972.
59. **Nečas, O. and Dyr, V.,** Die Lebensfähigkeit von Zellfragmenten der Hefen. Eigenschaften der regenerierenden Zellen, *Cesk. Biol.,* 6, 50, 1957.
60. **Nečas, O.,** Vitality of cell fragments in yeasts. IV. Relation of the nucleus to growth ability, *Folia Biol. (Praha),* 2, 29, 1956.
61. **Nečas, O.,** Multinuclear yeast cells, *Cesk. Biol.,* 5, 71, 1956.
62. **Havelková, M. and Křípalová, J.,** Independence of nuclear cycle on cell wall regeneration in *Schizosaccharomyces pombe* protoplasts, *Folia Microbiol. (Praha),* 25, 219, 1980.
63. **Svoboda, A. and Nečas, O.,** Morphogenesis during protoplast reversion in dimorphic yeasts, *Proc. 4th Int. Symp. on Yeast,* Part I, Vienna, 1974, p. D-13.
64. **Svoboda, A.,** Morphogenesis during regeneration and reversion of dimorphic yeasts, *Biologia (Bratislava),* 35, 157, 1980.
65. **Villanueva, J. R. and Garcia-Acha, I.,** The phenomenon of regeneration in fungal protoplasts, *Rev. Latinoam. Microbiol.,* 14, 49, 1972.
66. **Meissel, M., Medvedeva, G., Birjusova, V., Pomoschnikova, N., and Zvjagilskaja, R.,** Structure and development of normal and X-irradiated yeast protoplasts, in *Symposium über Hefeprotoplasten,* Müller, R., Ed., Akademie-Verlag, Berlin, 1967, 31.
67. **Svoboda, A.,** Regeneration ability of protoplasts of different yeast species, in *Symposium über Hefeprotoplasten,* Müller, R., Ed., Akademie-Verlag, Berlin, 1967, 31.
68. **Rost, K. and Venner, H.,** Wachstum von Hefeprotoplasten und ihre Reversion zu intakten Hefezellen, *Arch. Microbiol.,* 51, 130, 1965.
69. **Uruburu, F., Elorza, V., and Villanueva, J. R.,** The fine structure of yeast regenerating protoplasts, *J. Gen. Microbiol.,* 51, 195, 1968.

70. **Havelková, M. and Menšík, P.**, Ultrastructure of the cells and the protoplasts of *Lipomyces lipofer*, *Antonie van Leeuwenhoek J. Microbiol. Serol.*, 35B, 15, 1969.
71. **Deák, T. and Novák, E. K.**, Formation and regeneration of protoplasts in the genus *Candida*, in *Yeast Protoplasts*, Nečas, O. and Svoboda, A., Eds., J. E. Purkyně University, Brno, 1970, 217.
72. **Svihla, G., Schlenk, F., and Dainko, J. I.**, Sphaeroplasts of the yeast *Candida utilis*, *J. Bacteriol.*, 82, 808, 1961.
73. **Ota, F.**, Magnesium stimulated resynthesis of osmoresistant layer by *Candida albicans* sphaeroplasts, *Jpn. J. Microbiol.*, 16, 359, 1972.
74. **Zajkina, A. I. and Lapotyškina, A. I.**, Formation of protoplasts in the *Candida* yeasts, *Mikrobiologija*, 29, 1082, 1970.
75. **Müller, R.**, Mikroskopische Analyse der Protoplastenentstehung und Reversion, in *Symposium über Hefeprotoplasten*, Müller, R., Ed., Akademie-Verlag, Berlin, 1967, 37.
76. **Nečas, O.**, Die Regeneration von Zellfragmenten bei Hefen, *Biol. Zentralbl.*, 75, 308, 1956.
77. **Kopecká, M.**, Sporulation in protoplasts of the yeast, *Saccharomyces cerevisiae, J. Gen. Microbiol.*, 83, 171, 1974.
78. **Svoboda, A.**, Mating reaction in yeast protoplasts, *Arch. Microbiol.*, 110, 313, 1976.
79. **Indge, K. J.**, Metabolic lysis of yeast protoplasts, *J. Gen. Microbiol.*, 51, 433, 1968.
80. **Müller, R.**, Contribution to the problem of protoplast fusion, in *Yeast Protoplasts*, Nečas, O. and Svoboda, A., Eds, J. E. Purkyně University, Brno, 1970, 39.
81. **Kao, K. N. and Michayluk, M. R.**, A method for high frequency intergeneric fusion of plant protoplasts, *Planta*, 115, 355, 1974.
82. **Sipiczki, M. and Ferenczy, L.**, Protoplast fusion of *Schizosaccharomyces pombe* auxotrophic mutants of identical mating-type, *Mol. Gen. Genet.*, 151, 77, 1977.
83. **Ferenczy, L. and Maráz, A.**, Transfer of mitochondria by protoplast fusion in *Saccharomyces cerevisiae*, *Nature (London)*, 268, 524, 1977.
84. **Svoboda, A.**, Fusion of yeast protoplasts induced by polyethylene glycol, *J. Gen. Microbiol.*, 109, 169, 1978.
85. **Maráz, A., Kiss, M., and Ferenczy, L.**, Protoplast fusion in *Saccharomyces cerevisiae* strains of identical and opposite mating types, *FEMS Microbiol. Lett.*, 3, 319, 1978.
86. **Gunge, N. and Tamaru, A.**, Genetic analysis of products of protoplast fusion in *Saccharomyces cerevisiae*, *Jpn. J. Genet.*, 53, 41, 1978.
87. **Arima, K. and Takano, I.**, Multiple fusion of protoplasts in *Saccharomyces* yeasts, *Mol. Gen. Genet.*, 173, 271, 1979.
88. **Morgan, A. J., Heritage, J., and Whittaker, P. A.**, Protoplast fusion between petite and auxotrophic mutants of the petite-negative yeast, *Kluyveromyces lactis*, *Microbiol. Lett.*, 4, 103, 1977.
89. **Stahl, U.**, Zygote formation and recombination between like mating types in the yeast *Saccharomycopsis lipolytica* by protoplast fusion, *Mol. Gen. Genet.*, 160, 111, 1978.
90. **Fournier, P., Provost, A., Bourguignon, C., and Heslot, H.**, Recombination after protoplast fusion in the yeast *Candida tropicalis*, *Arch. Microbiol.*, 115, 143, 1977.
91. **Sipiczki, M. and Ferenczy, L.**, Fusion of *Rhodosporidium (Rhodotorula)* protoplasts, *FEMS Microbiol. Lett.*, 2, 203, 1977.
92. **Provost, A., Bourguignon, C., Fournier, P., Ribet, A. M., and Heslot, H.**, Intergeneric hybridization in yeast through protoplast fusion, *FEMS Microbiol. Lett.*, 3, 309, 1978.
93. **Ahkong, Q. F., Howell, J. I., Lucy, J. A., Safwat, F., Davey, M. R., and Cocking, E. C.**, Fusion of hen erythrocytes with yeast protoplasts induced by polyethylene glycol, *Nature (London), 255, 66, 1975.*
94. **Svoboda, A.**, Polyethylene glycol induced membrane fusion in yeast protoplasts, *Acta Histochem., Suppl.*, 23, 211, 1981.
95. **Sipiczki, M.**, Interspecific protoplast fusion in fission yeast, *Curr. Microbiol.*, 3, 37, 1979.
96. **Kucsera, J. and Ferenczy, L.**, Interspecific protoplast fusion between *Candida albicans* and *Candida tropicalis*, *Abstr. 5th Int. Protoplast Symp.*, Szeged, 1979, 52.
97. **Vallin, C., Kucsera, J., Klinner, U., and Ferenczy, L.**, Intergeneric protoplast fusion between *Candida tropicalis* and *Pichia guilliermondii*, *Abstr. 5th Int. Protoplast Symp.* Szeged, 1979, 59.

Chapter 7

AUTOLYSIS

W. N. Arnold

TABLE OF CONTENTS

I. INTRODUCTION

The phenomenon of autolysis has experimental and industrial applications. It is an appropriate subject for this volume in that events which collectively constitute autolysis include a profound effect on the cell envelope. Autolysis refers to the dissolution of cell substance after death of the cell and is due to the action of endogenous enzymes. Excluded within that context is the participation of polymer-modifying enzymes which are thought to be important in cell envelope expansion during growth of the living cell (see Volume II, Chapter 1). That is not to say, however, that some of the same β-glucanases might also participate in wall dissolution following death of the cell.

II. INITIATION

Autolysis begins immediately after cell death. Triggering events, be they natural or unnatural, include loss of membrane function (which results in deranged enzymes within the cell), and loss of chemical energy for the synthetic limb of turnover processes. By definition, only those treatments that spare at least some of the degradative enzymes can be considered autolysis initiators. For example, elevated temperatures initiate autolysis. On the other hand, cells that are plunged into boiling water quickly die, but do not undergo autolysis because their autolytic enzymes are inactivated.

The ideal, intentional initiator should result in immediate cell death so that membrane barriers are obliterated but at the same time this pretreatment should have no deleterious effect on key autolytic enzymes. All gradations between this ideal and the zero autolysis example given above are possible. If the goal of autolysis is to achieve release of a particular constituent from the yeast cells then an obvious further restraint is that the initiator not modify the compound of interest.

A. Aging

O'Sullivan and Thompson[1] allowed pressed brewery yeast to stand for 2 weeks at 15 to 20°C. After depletion of energy stores, the cells eventually died and underwent autolysis. But there is considerable likelihood of contamination by this simple method and it is no longer recommended for enzyme isolation.

A frequently encountered question with enzymes found in culture medium is to assign their origin to dead cells (autolysis) or to live cells (secretion). All too often the percentage of dead cells in experimental cultures is not monitored. This probably derives from experience with *Saccharomyces cerevisiae* in which less than 1% of cells die within normal cultural periods of 3 to 5 days. On the other hand, the author has observed that *Torulopsis glabrata* cultures grown under similar conditions will display 6% or more of dead cells after 2 days.[2] A slow-growing organism such as *Histoplama capsulatum* presents special problems that are not easily surmounted. The percentage of dead cells in experimental cultures is always significant, but can be minimized by using an inoculum from a preculture in exponential phase.[3] Dworschak and Wickerham[4] went to some trouble to prove that the unusual secretion of β-fructofuranosidase by *Sacch. uvarum* was by live cells. This brings up the general question of viability determination.

Viability is the ability to live and grow. Two approaches are used to judge the viability of yeast cells. One is based on dye exclusion and reduction which is actually indicative of active metabolism. We recommend the methylene blue method of Mills.[5] The other approach is based on the ability of a given cell to divide vegetatively and to give rise to a scorable colony on an agar plate. The former (staining) method usually yields a higher value for percent viability than the latter (colony formation) method. In part, this is due to the triviality that aggregated cells give rise to a single colony. But it also seems clear that some cells are able to meet the dye challenge, but are not able to carry out cell division. To estimate the percentage of cells in a given suspension that are undergoing autolysis, the dye exclusion method is probably the more meaningful.

B. Addition of Organic Solvents

A great number of organic compounds have been used at one time or another to initiate autolysis in yeasts. A review[6] on the early methods for β-fructofuranosidase isolation summarized the history of this subject. Hoppe-Seyler used ether for this purpose as early as 1871 and Salkowski employed chloroform through the turn of the century.[6] Subsequently, the most frequent agent was toluene, beginning with Hudson's studies on yeast enzymes.[7] Hudson used 2 to 3% toluene with approximately 50% suspensions of yeast, at room temperature, for 3 to 5 days. The following procedure is adapted[8] from Myrbäck and Schilling.[9]

A cake of compressed bakers' yeast is shaved all around with a spatula to remove the outermost layer of potentially dried cells; then 200 g of the remaining yeast cake is crumbled into 2 *l* of distilled water and brought into homogeneous suspension. The slurry is allowed to stand for about 15 min, decanted into bottles, and centrifuged at 3000 × g for 10 min. The supernate, together with a superficial rinse of the pellet with distilled water, are decanted to waste. The pellet is suspended in 1800 m*l* of 0.25 *M* sodium acetate buffer (pH 6.0), 100 m*l* toluene are added, and the autolysis conducted at 30°C. After 3 to 4 days, virtually all of the β-fructofuranosidase is released from the dead yeast cells without loss of total enzymatic activity. The rate of release of the above enzyme can be increased by addition of sulfhydryl compounds (cysteine or dithiothreitol) to the toluene-containing mixture.[10] The effect of the reducing compounds is to obliterate the otherwise 18 to 20 hr lag in β-fructofuranosidase release; the same effect can be obtained by using 2.5 mg/m*l* of *p*-toluenethiol alone as initiator.[10]

The susceptibility of enzymes to inactivation during autolysis is variable.[11] At one extreme the β-fructofuranosidase of *Sacch. cerevisiae* is quite resistant, but another periplasmic space enzyme, acid phosphatase, is susceptible to denaturation under similar organic solvents, pH, and temperature.[8,11] Again, the cryptic β-fructofuranosidase of *Sacch. rouxii* is expressed by a variety of organic solvents but suffers inactivation during a 2-day autolysis of cells.[12] In those cases where severe inactivation of specific enzymes has been noted it is usually not clear whether the prime cause is denaturation by the organic additive or susceptibility to proteolysis during autolysis. One case, the relative susceptibility of endogenous β-glucanases to a number of organic solvents, can be reproduced in a partially purified extract.[8]

When *Sacch. cerevisiae* cells are mixed with 0.1 vol 1-butanol at 30°C, there is no release of β-fructofuranosidase and about 25% of the activity is lost per day. Under similar conditions, *p*-dioxane or ethyl acetate elicit less than 5% release in 3 to 4 days, but there is good retention of total activity.[8] Based on these findings, and previous suggestions arising from the work by Myrbäck[13] and Burger et al.,[14,15] ethyl acetate exposure was chosen as a useful pretreatment which inactivates the autolytic enzymes and not β-fructofuranosidase.[8]

Ethyl acetate pretreatment is effected by adding 0.1 vol of solvent to a pellet of washed, wet cells. After kneading for several minutes at room temperature, the mass undergoes a pronounced change in viscosity. At this stage the cells already surrender a considerable fraction of their dry weight to the medium. If left at room temperature for 4 days then decreases in dry weight of cells (56%), total nitrogen (79%), and phospholipid (80%) accrue, but 85% of the β-fructofuranosidase activity remains and is sedimentable with the washed cell ghosts. This is a good model system for studying autolysis (see below).

C. Temperature Manipulation

Elevated temperatures obviously lead to premature death of yeast cells and autolysis commences. Vosti and Joslyn studied autolysis in *Sacch. cerevisiae*, at 53°C,[16] and in *Sacch. uvarum*, *Candida lipolytica* and a *Torulopsis* species (associated with flor sherry) at 45°C.[17] Changes in amino acid and total nitrogen concentrations were monitored and this work was also notable for emphasizing that the cell wall is not ruptured during these treatments. Under the light microscope at least, the wall is not greatly changed after 3 days of autolysis.[16] However, we now know that the porosity of the wall is increased by endogenous β-glucanase action (see below).

A method[10] for extracting enzymes from yeast at reduced temperatures involved first freezing cells at −195°C. For example, about 2700 g compressed yeast are crumbled into 15 *l* of liquid nitrogen and then kept at −20°C. The autolytic step is conducted at 5°C by mixing a portion of the frozen yeast with water, making slightly alkaline with ammonium hydroxide, and stirring slowly for 1 to 2 days. About 300 m*l* of cell-free extract is obtained after centrifugation of the mixture. There are idiopathic variations among commercial yeasts with respect to their ease of extraction by this method.[18a]

III. ENZYMOLOGY

A. Glucanases

Arnold[8] has shown that there are at least two β-glucanases (and possibly an aminopeptidase) that play a role in the autolytic dissolution of the cell wall in bakers' yeast. Both have endo action patterns, but one prefers β (1-6)-linked and the other β (1-3)-linked glucans as substrates. The autolytic enzymes were partially purified from cell-free extracts and tested against ethyl acetate-treated cells. The latter retain β-fructofuranosidase and the release of this marker to the medium provides a convenient monitor of wall modification.

Ethyl acetate inactivates the β-glucanases in question and this explains the lack of autolytic releases of β-fructofuranosidase from ethyl acetate-treated cells. On the other hand, the autolytic enzymes are stable at 30°C for several days in the presence of saturating amounts of toluene and are partly inactivated in this time frame by chloroform.[8]

The rate of release of marker enzyme from ethyl acetate-treated cells by purified glucanases is decreased by externally added glucans, but is not influenced by ethylenediaminetetraacetic acid (100 m*M*) or phenylmethylsulfonyl fluoride (2 m*M*). Active preparations were made from cells that had been pretreated with 0.1 *N* HCl. Such a treatment inactivated 96% of the β-fructofuranosidase of intact cells. Also, cell-free extracts of the β-glucanases were completely inactivated by similar acid treatment. These results show that some of the autolytic β-glucanases must be located internal to the plasma membrane, thanks to which they are protected from the acid pretreatment, and are still active when cell-free extracts are tested in the model system.[8]

The sulfhydryl status within the cell envelope seems to have an influence on autolysis. When ethyl acetate-treated cells were tested against a partially purified autolytic enzyme mixture, only 50% of the marker enzyme (β-fructofuranosidase) was released even after prolonged incubation. If cysteine, dithioerythritol, or mercaptoethanol were added along with the autolytic enzymes, then the release was complete. The effect of the reducing agents was apparently on some component(s) of the cell wall. Cells that were pretreated with reducing agent could be alkylated (with iodoacetamide) and they remained completely susceptible to β-glucanases without need of further additions. Carboxymethylcysteine was detected in acid hydrolyses of treated cells. Reduction and alkylation did not result in release of β-fructofuranosidase, but there was some loss of nitrogen (all dialysable).[8] That toluene-initiated autolysates go to completion would seem to indicate that the sulfhydryl status can be maintained in that case, although additions of cysteine, etc., stimulate.[10] Supposedly we are dealing here with the same or a related phenomenon associated with protoplast formation from older cells of *Sacch. cerevisiae* wherein it is necessary to add a sulfhydryl compound along with snail digestive juice (that contains β-glucanase).[19]

When ribonuclease or trypsin were added to ethyl acetate-treated yeast cells there was a rapid release of UV-absorbing materials into the medium. However, control material reached about the same level of depletion just by aging and slow leaching. It is possible that ethly acetate-treated cells still have residual ribonuclease and proteinases and that these activities are simply augmented by external additions. There is some preliminary evidence to suggest that the partially purified autolytic fraction contains an aminopeptidase; certainly some other enzymatic factor besides the purified β-glucanases is necessary for complete release of β-fructofuranosidase from reduced, ethyl acetate-treated cells.[8]

B. Other Enzymes

There is a plethora of proteinases in yeast and they play important regulatory roles in living cells.[20] In addition, it is clear that various enzymes are subject to modification during isolation by autolysis.[21-23] Also, proteinases have been invoked in the overall process of cell wall dissolution. In this last regard, it is proven that the β-glucanases play the major role, but studies on the ethyl acetate-treated cell have suggested that nonyeast proteinases can also effect release of β-fructofuranosidase.

Willstätter and Grassman[24] were convinced that rather large amounts of trypsin followed by papain could accelerate the release of β-fructofuranosidase from cells treated 1 day with ethly acetate. However, it was only after the papain treatment that the marker enzyme became detached from the treated cells. A series of papers by Myrbäck and colleagues followed up this work and showed that pancreatin, which is rich in trypsin, amylase and lipase, had no direct action on β-fructofuranosidase release.[25] On the other hand, they could get respectable release with 1% papain and 0.5% cysteine,[25] or malt extract,[26] which contains endo-(1-3)-β-glucanase.[27]

Bacon and colleagues embraced the utility of the ethyl acetate–treated cell and showed that commercial papain would liberate 51% of the β-fructofuranosidase from 20-hr-treated ethyl acetate cells,[14] compared with disintegration in a Hughes press (88%),[15] or after treatment with snail gut enzymes (96%).[15] However, controls released 15% marker in 2 weeks and a more purified papain released only 16% in the same time frame.[14] The last result, together with the large amounts of papain usually required by others, suggested the participation of some other enzyme in crude additions. Wilson[28] has demonstrated that some commercial preparations of papain contain an endo-β-(1-3)-glucanase with pH optimum of 5.0. Although this has not been followed up in the yeast application, it should now be relatively straightforward to explore.

Previous workers invariably used 1-day pretreatments with ethyl acetate, but this is insufficient for complete inactivation of the autolytic enzymes. Accordingly, the author has found that cells pretreated for 1 to 4 days with ethyl acetate, washed, and held under toluene, subsequently released 66%, 8%, 5%, or 3% of their β-fructofuranosidase in the next 2 weeks. For this reason longer treatments in the preparation of the ethyl acetate model is recommended. With a 20% suspension of 4-day treated cells, the author obtained quantitative release of β-fructofuranosidase after 4 days in the presence of 0.35 mg/m*l* papain (from Sigma) and 1 mg/m*l* cysteine; and less than 1% release in the absence of additives. The author also observed 72% release with 0.63 mg/m*l* pronase, but this was accompanied by 12% loss of total activity. Unfortunately, pronase (ex. *Steptomyces griseus*) is a mixture of proteinases.[29]

The effect of proteinases on the ethyl acetate-treated cells is still not understood. If the proteinases are able to elicit release of β-fructofuranosidase, then two possibilities are suggested. Either we are dealing with a structural protein in the wall matrix (and this could be related to the sulfhydryl effects mentioned earlier), or the proteinases may be generating a modified β-fructofuranosidase that is more easily eluted. Both hypotheses should be tested and it would also be of interest to monitor concurrent ultrastructural changes in the ethyl acetate-treated cell.

The participation of lipases and phospholipases in wall dissolution is not indicated. However, any lipid-constituent that is isolated from yeast by autolysis will be subject to more or less degradation depending upon conditions. The preparation of active dried bakers' yeast does not involve autolysis by the definition of this chapter, but some of the same enzymes[30] that interfere during the dehydration and then later the rehydration of the plasma membrane[30,31] may also play a role in the early events of natural autolysis. The action of lecithinase C is most important in this respect.[30] It is worth pointing out that 0.1% of several detergents did not release β-fructofuranosidase from the ethyl acetate-treated cell;[31a] these detergents included sodium deoxycholate, Triton X-100®, digitonin, hexadecyltrimethylammonium bromide, cetylpyridinium chloride, and cetylmethylbenzylammonium chloride. Incidentally, the last three treatments did result in 55, 50, and 24% (respectively) inactivation of marker enzyme.

IV. PHYSICAL ASPECTS

Death of yeast cells results in immediate loss of turgor, and in the case of ethyl acetate treatment, the packed cell volume of cells decreases by 48%. This effect is a manifestation of plasma membrane dysfunction if not disruption. Arnold and Lacy[32] showed that 3% glutaraldehyde fixation can preserve cells at about 97% of their native volume, but the common use of buffers during fixation leads to larger volume changes (e.g., up to 32% decrease). Incidentally, the mean cell volume for dried and reconstituted yeast is only 65% of that before drying.[33]

To the best of my knowledge, the autolyzing cell has not been subjected to ultrastructural study so that putative changes in the cell wall have not been monitored by electron microscopy. During the first several days of autolysis, it is generally assumed that pores are generated in the cell wall, rather than whole walls being digested; this follows from observations of light microscopy and companion estimates of cell ghost number.[8,17]

Arnold[8] attempted to get some idea of average pore size during toluene-induced autolysis. An experiment was set up to test the progressive susceptibility of autolyzing cells to washing on the centrifuge. It was argued that as the β-fructofuranosidase was set free from the cells, so also should the autolytic enzymes be released and subject to removal by washing. Until 8% of the marker was released, washing had no effect on the subsequent autolysis rate. But after 20% of the marker was released, cells had their subsequent autolytic rate very significantly lowered by washing. This indicates that a significant part of the autolytic system was in the medium at that point. It is known that there is very little effect of cell concentration on autolysis rate and this suggests that each cell does its own thing during autolysis. Thus, the possibility that 20% of the cells had released all of their β-fructofuranosidase seems to be ruled out because subsequent autolytic rates would then have been normal. Because yeast β-fructofuranosidase is polydisperse in polysaccharide content,[34,35] it would now be of interest to compare the ranges of molecular weights between early and late autolytic products. If the expanding pore hypothesis is correct, one would expect a smaller, average molecular weight for the enzyme molecules that emerge first. Incidentally, this might rationalize some of the outstanding differences in polysaccharide content of autolytic preparations of β-fructofuranosidase in the early literature.

V. LABORATORY AND INDUSTRIAL APPLICATIONS

The isolation of enzymes from yeast via autolysis has been largely superseded by mechanical disruption methods (see Volume I, Chapter 3). This is, in part, due to the sensitivity of some enzymes to the autolytic conditions and, more importantly, the general fear that covalent modifications may be suffered during prolonged contact with endogenous proteolytic enzymes. Modifications of this sort have been documented; in other cases, and notably β-fructofuranosidase, there has been no indication that autolytic release is accompanied by limited proteolysis (see Volume II, Chapter 1).

If an enzyme is to be used in the assay of a metabolite, either directly or in coupled enzyme assay, then it is probably of no import that the analytic agent has been slightly modified through autolysis as long as the specificity is not altered. In fact, one assumes that several of the yeast enzymes that are commercially available are still isolated by autolysis. The main advantages of the autolytic method are that it does not demand sophisticated equipment and, although somewhat longer overall, it is very short on operator contact time.

Other industrial applications of autolysis include preparations for food-addition, food extension, and texturizing. The first is enriched in peptides and proteins, the last two products are predominantly glucans. These preparations involve proprietary methods that are held under various patents, but are based on variations of the methods already indicated. Another application of autolyzed yeast is in the preparation of microbiological media which is now standardized in dehydrated form.

For culture media, yeast autolyzate is made by holding equal parts of yeast and water at 55°C for 3 days; briefly boiling the suspension, and then filtering until clear. Commercial preparations, in the U.S. at least, are marketed under the name "yeast extract", i.e., the water soluble portion of autolyzed yeast. This should not be confused with classical yeast extract which is an infusion prepared by suspending 1 part bakers' yeast with 5 parts water; autoclaving at 15-lb pressure for 15 min; and filtering until clear. As pointed out by Bouthilet et al.[36] classical yeast extract may be rich in trehalose whereas autolysate prepared by the above method is not. This can have taxonomic implications,[36] and other differences in the two preparations would not be unexpected. In this laboratory we have observed that *Sacch. rouxii* cultures occasionally display a high proportion of elongated cells, and the propensity for abnormal cells seems to depend on the commercial source and age of the yeast extracts used in our complex medium.

Yeast autolysis can have a positive influence on the quality of some wines because autolytic products encourage secondary malo-lactic fermentations, i.e., in the manufacture of flor sherries.[37] On the other hand, spoilage may result if an undesirable bacterial infection occurs.[38] The loss of leavening power in bakers' yeast during storage and changes in brewers' yeast during pitching are clearly complex phenomena but almost certainly involve some degree of autolysis.

VI. CONCLUDING REMARK

The biochemical and biophysical events associated with yeast autolysis have not attracted a great deal of study. This is because related problems are in large part industrial and are solved empirically. However, the release of enzymes from autolyzing cells or from the ethyl acetate treated model may be worth further study and could provide additional insight into the disposition of polymers within the cell wall. One deficiency at present is ultrastructural description.

REFERENCES

1. **O'Sullivan, C. and Thompson, F. W.,** Invertase: a contribution to the history of an enzyme or unorganized ferment, *J. Chem. Soc.*, 57, 834, 1890.
2. **Arnold, W. N.,** Trehalose assimilation and turnover in *Torulopsis glabrata, Curr. Microbiol.*, 2, 109, 1979.
3. **Reca, M. E. and Campbell, C. C.,** Growth curves with yeast phase of *Histoplasma capsulatum, Sabouraudia*, 5, 267, 1967.
4. **Dworschack, R. G. and Wickerham, L. J.,** Production of extracellular invertase by the yeast *Saccharomyces uvarum* NRRL Y-972, *Arch. Biochem. Biophys.*, 76, 449, 1958.
5. **Mills, D. R.,** Differential staining of living and dead cells, *Food Res.*, 6, 361, 1941.
6. **Neuberg, C. and Roberts, I. S.,** Invertase, *Sugar Res. Found. Sci. Rep. Ser., 4,* 1946.
7. **Hudson, C. S.,** The inversion of sucrose by invertase VII. An improved method for preparing strong invertase solutions from top or bottom yeasts, *J. Am. Chem. Soc.*, 36, 1566, 1914.
8. **Arnold, W. N.,** The structure of the yeast cell wall. Solubilization of a marker enzyme, β-fructofuranosidase, by the autolytic enzyme system, *J. Biol. Chem.*, 247, 1161, 1972.
9. **Myrbäck, K. and Schilling, W.,** Studies on yeast β-fructofuranosidase. XVI. Partial purification, *Enzymologia*, 29, 306, 1965.
10. **Arnold, W. N.,** *p*-Toluenethiol as an initiator of autolysis in bakers' yeast, *J. Bacteriol*, 109, 949, 1972.
11. **Arnold, W. N.,** Location of acid phosphatase and β-fructofuranosidase within yeast cell envelopes, *J. Bacteriol.*, 112, 1346, 1972.
12. **Arnold, W. N.,** Expression of cryptic β-fructofuranosidase in *Saccharomyces rouxii, J. Bacteriol.*, 120, 886, 1974.
13. **Myrbäck, K.,** Studies on yeast invertase (Saccharase) of bakers' yeast, *Arch. Biochem. Biophys.*, 69, 138, 1957.
14. **Burger, M., Bacon, E. E., and Bacon, J. S. D.,** Some observations on the form and location of invertase in the yeast cell, *Biochem. J.*, 78, 504, 1961.
15. **Burger, M., Bacon, E. E., and Bacon, J. S. D.,** Liberation of invertase from disintegrated yeast cells, *nature (London),* 182, 1508, 1958.
16. **Vosti, D. C. and Joslyn, M. A.,** Autolysis of bakers' yeast, *Appl. Microbiol.*, 2, 70, 1954.
17. **Vosti, D. C. and Joslyn, M. A.,** Autolysis of several pure culture yeasts, *Appl. Microbiol.*, 2, 79, 1954.
18. **Black, S. and Wright, N. G.,** β-Aspartokinase and β-aspartyl phosphate, *J. Biol. Chem.*, 213, 27, 1955.
18a. **Black, S.,** personal communication.
19. **Duell, E. A., Inoue, S., and Utter, M. F.,** Isolation and properties of intact mitochondria from spheroplasts of yeast, *J. Bacteriol.*, 88, 1762, 1964.
20. **Hansen, R. J., Switzer, R. L., Hinze, H., and Holzer, H.,** Effects of glucose and nitrogen source on the levels of proteinases, peptidases and proteinase inhibitors in yeast, *Biochim. Biophys. Acta*, 496, 103, 1977.
21. **Juni, E. and Heym, G. A.,** Properties of yeast pyruvate decarboxylase and their modification by proteolytic enzymes. II. Selective alteration by yeast proteases, *Arch. Biochem. Biophys.*, 127, 89, 1968.
22. **Sasaki, R., Sugimoto, E., and Chiba, H.,** Yeast phospoglyceric acid mutase modifying enzyme, *Arch. Biochem. Biophys.*, 115, 53, 1966.
23. **Kaji, A.,** Partial proteolytic digestion of yeast hexokinase and its relation to multiple forms of the enzyme, *Arch. Biochem. Biophys.*, 112, 54, 1965.
24. **Willstätter, R. and Grassman, W.,** Zur Freilegung des Invertins aus der Hefe, *Biochem. Z.*, 203, 308, 1928.
25. **Myrbäck, K. and Willstaedt, E.,** Studies on yeast invertase (Saccharase). Localization of the enzyme in the cell and its liberation, *Arkiv Kemi*, 8, 367, 1955.
26. **Thorsell, W. and Myrbäck, K.,** Insoluble saccharase in bakers' yeast, *Arkiv Kemi*, 3, 323, 1951.
27. **Manners, D. J. and Wilson, G.,** Purification and properties of an endo-(1-3)-β-D-glucanase from malted barley, *Carbohyd. Res.*, 37, 9, 1974.
28. **Wilson, G.,** Some properties of an endo-1, 3-glucanase from papain, *Biochem. Soc. Trans.*, 2, 115, 1974.
29. **Trop, M. and Birk, Y.,** The specificity of proteinases from *Streptomyces griseus* (Pronase), *Biochem. J.*, 116, 19, 1970.
30. **Harrison, J. S. and Trevelyan, W. E.,** Phospholipid breakdown in bakers' yeast during drying, *Nature (London),* 200, 1189, 1963.
31. **Herrera, T., Peterson, W. H., Cooper, E. J., and Peppler, H. J.,** Loss of cell constituents on reconstitution of active dried yeast, *Arch. Biochem. Biophys.*, 63, 131, 1956.
31a. **Arnold, W. N.,** Unpublished data.

32. **Arnold, W. N. and Lacy, J. S.,** Permeability of the cell envelope and osmotic behavior in *Saccharomyces cerevisiae, J. Bacteriol.*, 131, 564, 1977.
33. **Ebbutt, L. I. K.,** The relationship between activity and cell-wall permeability in dried bakers' yeast, *J. Gen. Microbiol.*, 25, 87, 1961.
34. **Arnold, W. N.,** Heat inactivation kinetics of yeast β-fructofuranosidase. A polydisperse system, *Biochim. Biophys. Acta,* 178, 347, 1969.
35. **Arnold, W. N.,** Chromatography of a polydisperse enzyme, yeast β-fructofuranosidase, on DEAE-cellulose, *Anal. Lett.*, 2, 547, 1969.
36. **Bouthilet, R. J., Neilson, N. E., Mrak, E. M., and Phaff, H. J.,** The fermentation of trehalose by yeasts and its taxonomic implications, *J. Gen. Microbiol.*, 3, 282, 1949.
37. **Fornachon, J. C. M.,** The occurrence of malolactic fermentation in Australian wines, *Aust. J. Appl. Sci.*, 8, 120, 1957.
38. **Fornachon, J. C. M., Douglas, H. C., and Vaughn, R. H.,** *Lactobacillus trichodes* nov. spec., a bacterium causing spoilage in appetizer and dessert wines, *Hilgardia,* 19, 129, 1949.

Chapter 8

VEGETATIVE ULTRASTRUCTURE

R. G. Garrison

TABLE OF CONTENTS

I. INTRODUCTION

The yeast cell envelope includes the plasma membrane, the cell wall, and (in some cases) a slime layer (see Volume I, Chapter 1) all of which are anatomically distinct. The plasma membrane and the inner aspect of the cell wall circumscribe the periplasmic space which is an important region of the envelope but can be anatomically nebulous after some staining regimens. This chapter is devoted to the ultrastructure of the yeast cell envelope; documentation that is intrinsically important as well as being complementary to many biochemical and biophysical investigations. Background information on yeast cell cytology is given in the review by Matile et al.[1] In Volume I, Chapter 2 we illustrated yeast morphology and emphasized cell shape and surface topography; the focus here is on finer detail within the envelope.

Most of our information on cell envelope anatomy has come from electron microscopy of ultrathin sections of fixed cells, and of replicas cast on the surface of freeze-fractured cells. It is acknowledged at the outset that there is room for improvement in some preparatory techniques, especially those connected with fixation and staining of ultrathin sections. One is struck by the narrow applicability of so-called routine methods among species, or even within a species according to cell age or growth conditions. In general, electron micrographs of thin sections of yeasts have not matched the quality of images obtained with higher eukaryotic cells. Notwithstanding these deficiencies current methodology provides an indispensable means of visualizing components of the cell envelope.

A detailed treatment of the principles and techniques of electron microscopy is beyond the scope of the present chapter, but certain fundamental techniques are briefly discussed for the sake of those entering the field. Detailed instructions for the preparation of fixed and embedded materials for transmission electron microscopy are included in a number of manuals.[2-5] Greenhalgh and Evans,[6] and Kopp[7] emphasize basic techniques for fungal materials. Comprehensive and detailed information is available on freeze-fracture, freeze-etching, and replication.[8-12]

II. BASIC APPROACHES

Thin sectioning is probably the best approach for any initial study of cell structure. For electron microscopy this entails ultrathin sectioning of properly fixed and stained material, and requires the greatest effort in all aspects of specimen preparation. The question of possible preparatory artifacts arising from chemical fixation and dehydration is always with us. But the application of two or more processing regimens greatly increases the confidence in the interpretations of structures.

The freeze-fracture approach is widely employed and has some inherent advantages, particularly in the detailed examination of the plasma membrane. The examined object is actually a carbon and metal replica of a surface. The steps include freezing the yeast suspension, fracturing the block, etching the exposed surface by sublimation of ice, carbon coating, and platinum shadowing. Since only the replica is examined in the electron microscope there is no radiation damage to contend with as there is in the case of organic material. Since fixation and dehydration are not involved, some potential artifacts are minimized by the freeze-fracture approach. Yet this technique is not without problems; freezing artifacts may occur, and interpretations of images are sometimes difficult.

Technical improvements in scanning electron microscopes (SEM) accompanied by operator skill, have resulted in some exquisite micrographs in recent publications. Cells are fixed and dehydrated through an alcohol series and then subjected to critical-point drying with freon. The dried material is attached to a conductive stub with tape and the whole surface is sputter coated with a fine film of gold. It is the gold-coated specimen that is subsequently photographed in a SEM or scanning transmission electron microscope (STEM). The utility of the method is confined to topological studies of whole cells or protoplasts. Some distortion of delicate structures is almost inevitable but can be minimized by critical-point drying.

It is clear that each of the three approaches mentioned above has inherent advantages and disadvantages, but that each approach can make a valuable contribution to the morphology and cytology of envelopes. Complementarity is the key and, given the resources, an intelligent application of all three methods will provide a well-rounded anatomical description.

A. Fixation

1. Permanganate

Although some microscopists consider permanganate to be a capricious fixative at best, in some cases it may be the only chemical fixative to adequately preserve membranous structure and to delineate unit membranes. Potassium permanganate is most widely used although the barium salt has been recommended for enhancement.[13] Lithium, sodium, calcium, or zinc salt has no demonstrable advantage.[14] Electron micrographs of yeast cells fixed in permanganate usually display defined cytomembranes against a poorly differentiated background. Ribosomes are destroyed. Optimal conditions vary with the yeast species but the parameters include concentration of fixative, presentation time, temperature, and extent of washing. The author has most often employed unbuffered 1% (wt/vol) potassium permanganate for 30 min in an ice-water bath, following a variation of Mollenhauer.[15] Figures 1A and 1B show thin sections of *Saccharomyces cerevisiae* fixed in potassium and barium permanganate respectively. The latter shows more differentiation within bud scars. Representative examples rather than "best" results were intentionally selected in the plates of Figure 1 in order to convey general expectations for newcomers to the field.

An unpublished method by Arnold uses washed cells that have been incorporated into agar cylinders. The tiny pellets are exposed to 5% (wt/vol) $KMnO_4$ for only 2 min and then passed quickly through 25% ethanol into 50% ethanol. The specimens are further dehydrated through a typical alcohol series. Processing of samples is very rapid, and no centrifugation steps are involved. We find that cytoplasmic constituents are better preserved and good differentiation of cell envelopes is possible. The method needs further testing. Supposedly the rapid dehydration step avoids post permanganate extraction normally encountered in washing.

2. Glutaraldehyde—Osmium Tetroxide

Our routine method of fixation employs 3% glutaraldehyde as primary fixative followed by 1% osmium tetroxide as postfixative and stainer. A highly purified glutaraldehyde (designated electron microscopic grade by some distributors) is recommended. The fixed samples are washed with 0.9% NaCl after exposure to each fixative.

A great number of variables can be considered as more or less influential. These include the buffer employed (see also Volume I, Chapter 3 for biophysical evidence with respect to shrinkage), tonicity, temperature, fortification with Mg^{++}, and washing regimens. Buffers in vogue include s-collidine (2,4,6-trimethyl pyridine), cacodylate (dimethylarsinate), and phosphate; usually employed in final concentrations of 0.1 *M* and with chloride and sodium as the counterions. This author usually uses collidine buffer at pH 6.8; most literature references mention a pH close to neutrality; and

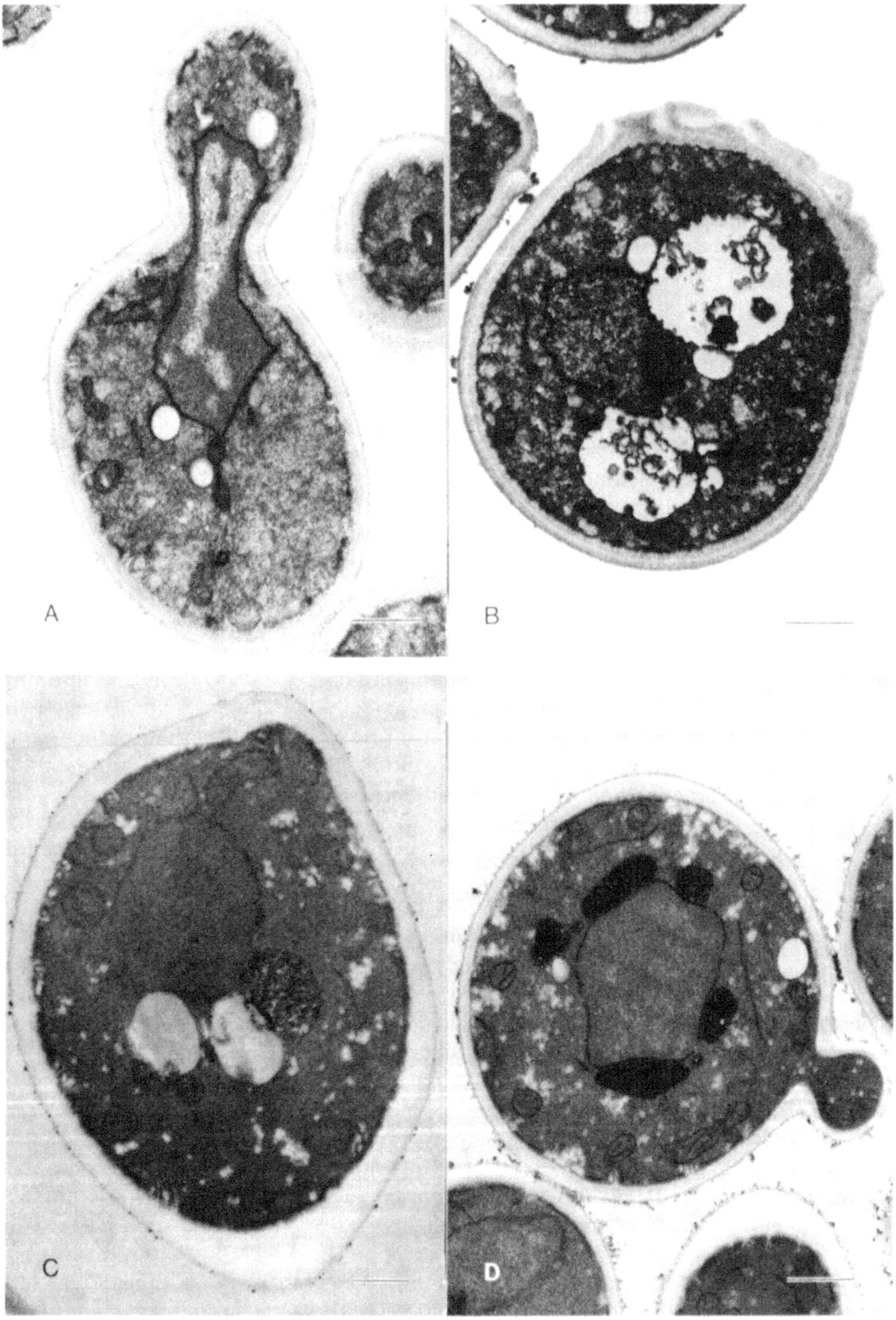

FIGURE 1. Effect of different fixatives on the resulting ultrastructure of *Saccharomyces cerevisiae*. All ultrathin sections were stained with uranyl acetate and lead citrate before examination. (A) 1% potassium permanganate, 30 min, 4°C; (B) 1% barium permanganate, 30 min, 4°C; (C) 3% glutaraldehyde, 24 hr, 4°C followed by 1% osmium tetroxide, 20 hr, 4°C; and (D) 3% glutaraldehyde, 24 hr, 4°C followed by 1% potassium permanganate, 30 min, 4°C. All bars equal 0.5 μm.

usually fixes in glutaraldehyde overnight in a refrigerator (about 4°C) but other workers use 1 to 4 hr at room temperature. We find no single regimen for glutaraldehyde—osmium fixation to be consistently superior for all yeast species.

It is generally recognized that fixation of most yeasts in glutaraldehyde—osmium tetroxide does not yield as successful a result as typically encountered with higher eukaryotic cells. A representative example with *Sacch. cerevisiae* is given in Figure 1C and shows rather poor visualization of the cell envelope and cytoplasmic structures. These difficulties are most likely due to poor penetration of the yeast cell wall by osmium tetroxide because much better cytological detail is the norm for protoplasts and spheroplasts.

Permanganate may be substituted for osmium tetroxide as post fixative (Figure 1D), in which case the membranous detail is usually improved. In the same context tannic acid has been used as a mordant,[16] and see also Figure 3A. The inclusion of tannic acid in the fixation regimen with glutaraldehyde—osmium tetroxide holds promise and needs to be explored on many additional species of yeasts.

3. Other Fixatives

Tris-(1-aziridinyl) phosphine oxide (TAPO), used in conjunction with acrolein and osmium tetroxide, has been employed to enhance contrast in the cell envelopes of *Candida albicans*,[17] *Cryptococcus neoformans*,[18] and *Sacch. cerevisiae*.[19] Results have been variable in my hands and the mechanism by which TAPO reacts with cell constituents is unknown. The compound is carcinogenic.

Some workers find that soaking fixed specimens in uranyl acetate enhances cytological detail.[20] With respect to yeasts there is little data available on more specialized fixatives such as chromium and mercury salts and phosphotungstic acid, but it is doubtful that they possess any distinct advantages over those already mentioned.

B. Dehydration and Embedment

Fixed yeast cells are usually incorporated as a thick suspension into 2% agar to facilitate subsequent handling.[21] Small cubes or cylinders are recommended in order to increase the rate of exchange of solvents. Other supports include fibrin clots,[22] nucleohistone clots,[23] and denatured egg or serum albumins.[24,25]

Schedules of dehydration and methods of infiltration depend largely on the choice of embedment. Detailed descriptions of these procedures are often given by manufacturers as well as by standard texts. Useful embedments include epoxy resins (e.g., Epon®, Araldite®, Maraglas®, and Durcupan®), a few methacrylates (*n*-butyl and *n*-methyl derivatives), and a polystyrene plastic (Vestopal®). Epons containing nadic methyl anhydride are incompatible with permanganate fixed material.[16,26,27] The author recommends the Epon® mix of Spurr[28] for yeasts because of its low viscosity.

C. Sectioning and Staining

Sections of about 70 to 80 nm in thickness are cut with glass or diamond knives and placed on 300-mesh copper grids. At times a supporting film of Formavar or carbon may be necessary. Gold grids are required if certain stains that are incompatible with copper are anticipated.

Staining of cells before embedment can be achieved by soaking in an aqueous, saturated solution of uranyl magnesium acetate.[14] But for the most part staining is performed on thin sections and involves successive treatments with 2% aqueous or alcoholic solutions, of uranyl acetate followed by lead tartrate or lead citrate.

III. THE CELL ENVELOPE

The problem of cell wall impermeability towards fixatives, especially osmium tetroxide, has already been mentioned. Partial removal of the cell wall by mechanical or enzymic means can ameliorate this situation but in the context of envelope ultrastructure these devices can hardly be recommended. (This subject is also discussed in Volume I, Chapter 3.) In the older literature the yeast cell envelope was usually poorly stained. However, with modern techniques various features can be accentuated and these will now be discussed in turn.

A. Fimbriae

Fine, hair-like appendages have been described on the walls of certain ascomycetous and heterobasidiomycetous fungi.[29-31] They are observed in the electron microscope after negative staining with phosphotungstate, shadow casting with tungsten oxide, or freeze-etching. It has been postulated[31] that fimbriae are involved in sexual conjugation and in flocculation. Thus, it is claimed that flocculent strains of *Sacch. cerevisiae* and *Sacch. uvarum* are endowed with more fimbriae than are nonflocculent strains.

B. Slime Layer

A few species of yeast generate a conspicuous slime layer as the outermost feature of their cell envelopes. Slime layers or capsules are largely polysaccharide[32] and their chemistry is discussed in Volume I, Chapters 5 and 6, Volume II, Chapter 2. After conventional fixation and staining of cells the slime layers may be translucent but alternative enhancement procedures are available. These include the use of ruthenium red,[33,34] plant lectins conjugated with ferritin or iron dextrans,[35] and iron colloids.[36] The capsule of *Cr. neoformans* is demonstrable in freeze-fractured specimens but only after deep etching.[37]

The slime layer of some species seems to contain a matrix of delicate microfilaments. In *Cr. neoformans* these microfilaments are only 3 to 4 nm in diameter, are coiled and intertwined, but are said to contain only polysaccharides, on the basis of staining properties at least.[38] Capsular microfilaments are visible in the specimen of *Cr. neoformans* shown in Figure 2. They are heavily stained by the periodate procedure of Thiéry[39] (Figure 2B). Microfilaments probably contribute to the stability of slime layers; any other function has yet to be demonstrated.

While most yeasts are devoid of a true slime layer several species are endowed with an outermost zone of the cell wall which is distinctly microfibrillar in nature. Examples are *Candida albicans* (Figures 3A and 4A), *Kluyveromyces polysporus* (Figure 3B), *Pichia membranaefaciens* (Figure 4B), *Hanseniaspora osmophila* (Figure 4C), and *C. tropicalis* (Figure 5). The microfibrillar zone seems to be diminished or absent in buds of some species which exhibit a pronounced zone in mature cells.[14]

In mature cells of *C. albicans* the microfibrillar zone ranges in thickness from 55 to 80 nm and is composed of anastomosing fibrils. The latter have increased electron opacity compared with the cell wall. Granules that are about 3 nm thick are embedded in the filament network.[17] Tannic acid mordant seems to intensify the staining of filaments[16] as shown in Figure 3A. Because yeast cells of *C. albicans* are strongly agglutinated by concanavalin A it has been suggested that the microfibrillar zone may be rich in mannan.[40] The microfibrillar zone may play a role in adherence, a subject that has received little attention to date but is supposedly important in the invasion process of pathogenic yeasts.

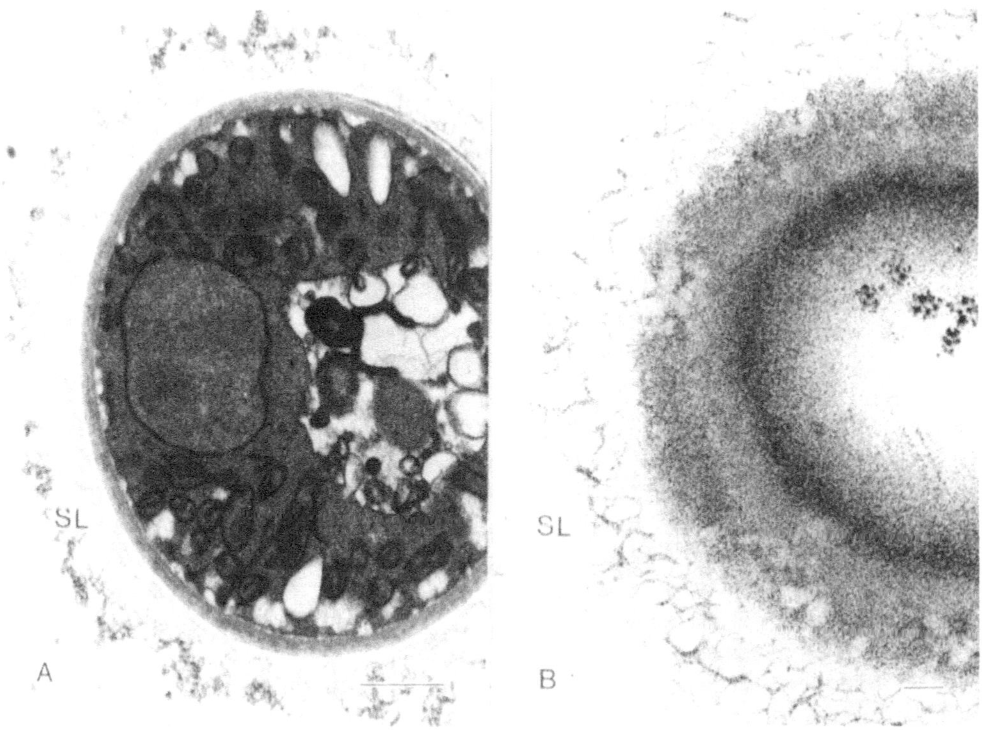

FIGURE 2. Transmission electron micrographs of *Cryptococcus neoformans*. Note the slime layers (SL). (A) Permanganate fixation, bar equals 0.5 μm; (B) periodate procedure of Thiéry. Note deposition of silver grains on filaments. Bar equals 0.1 μm.

C. The Cell Wall

The chemistry of the yeast cell wall is discussed in Volume I, Chapters 5, 6, and 7, and Volume II, Chapter 2. It is generally agreed that composition cannot be rationalized with ultrastructure at present but broad outlines are possible (see also Volume I, Chapter 5). Accordingly, the whole subject of substructure within cell walls is very nebulous at present and the plethora of claims for "layers" within walls cannot be interpreted on a chemical basis. The retention of electron opaque materials to varying degrees within walls could be due to physical parameters rather than chemical differences. Staining with osmium tetroxide, uranyl acetate, and lead citrate revealed irregular thickening in the walls of *Sporobolomyces roseus* (Figure 6A). An extreme example of alternate dark and light bands due to permanganate is shown by *Sterigmatomyces halophilus* (Figure 6B). It has been proposed that the structure of the cell wall differs among yeasts with different taxonomic affinities.[41]

Cassone and associates[17-19] distinguish many regions within the cell walls of *Sacch. cerevisiae, C. albicans*, and *Cr. neoformans* using the TAPO -acrolein fixatives. Although their technique substantially increases contrast within the walls of some specimens we find the procedure somewhat variable with *Saccharomyces rouxii* cells.

A number of investigators have attempted to learn something about the disposition of wall components by conducting ultrastructural studies on partially digested cells.[42-44] The enzyme sources have been pronase and snail gut juices—unfortunately both contain a multiplicity of enzymes (and see Volume II, Chapters 1 and 5). It is not unreasonable to expect that the use of more defined enzyme preparations together with the possible

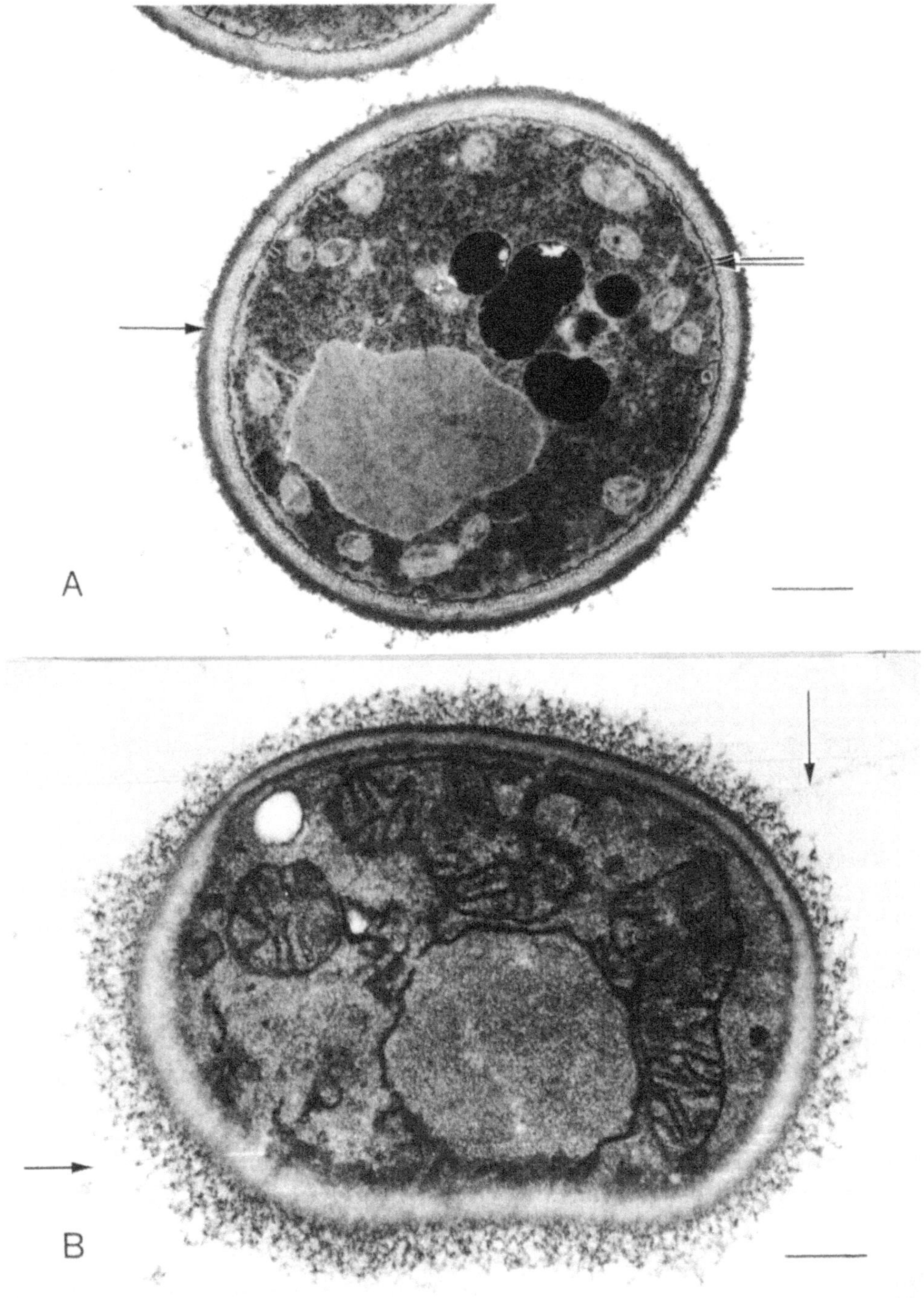

FIGURE 3. Cells displaying microfibrils. (A) *Candida albicans*, glutaraldehyde—osmium tetroxide in conjunction with tannic acid as mordant. Note outermost region of microfibrillar material (arrow). Periplasmic space is heavily stained (double arrow). Bars equal 0.5 μm. (Courtesy of M. Persi.) (B) *Kluyveromyces polysporus*. Note the elongated microfibrils (arrows). (Permanganate.) Bar equals 0.25 μm.

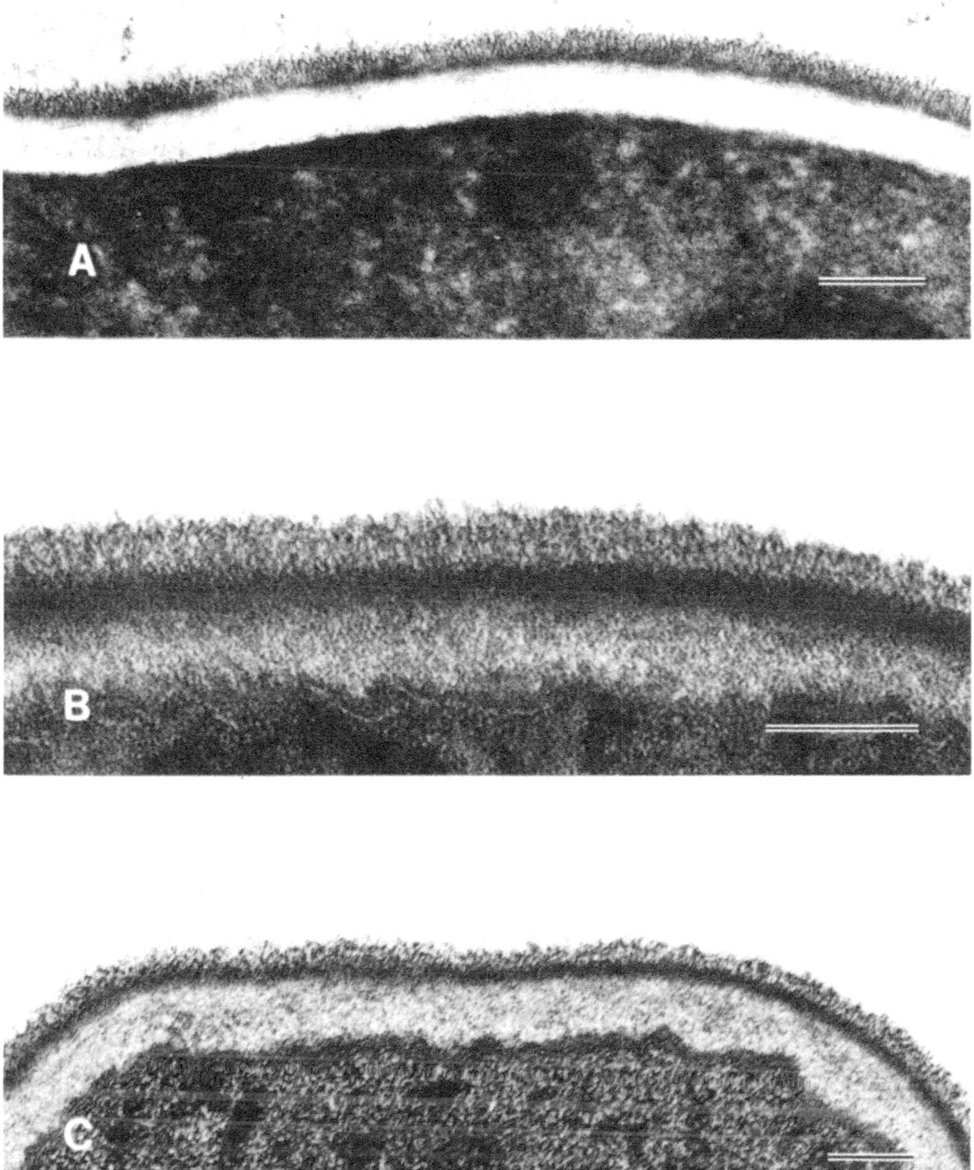

FIGURE 4. Cells displaying microfibrils. (A) *Candida albicans* (Permanganate.); (B) *Pichia membranaefaciens* (Glutaraldehyde—osmium tetroxide.); and (C) *Hanseniaspora osmophila* (Permanganate.) All bars equal 0.1 μm.

application of electron opaque and specific markers (antibodies and lectins) might yield some insight into this area in the near future.

Estimates of cell wall thickness from electron micrographs should be interpreted with caution. There is the trivial problem of plane of section, but several physiological and environmental factors, either singly or in combination, can influence cell wall thickness to the extent that this criterion has very little diagnostic value. According to Hagedorn[45]

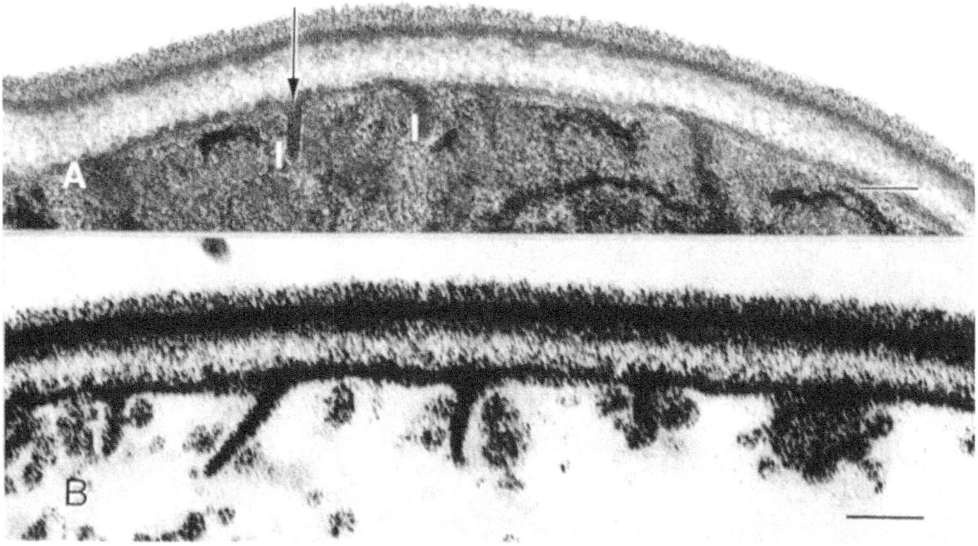

FIGURE 5. Transmission electron micrographs of *Candida tropicalis*. (A) Note the outer fibrillar zone, the dark and light staining regions of the cell wall, and the invaginations (I) in the plasma membrane. The periplasmic space (arrow) includes invaginations into the plasma membrane. (Permanganate.) (B) Periodate procedure of Thiéry. Bars equal 0.1 μm.

the cell wall of *Sacch. cerevisiae* can be 100- to 400-nm thick depending on cellular age. Oxygen tension as well as limitations in the medium of carbon, nitrogen, and phosphorus may affect cell wall thickness and appearance[46] together with mechanical properties (Volume I, Chapter 3). Trace element deficiencies may have effects on the wall structure of fungi; a case in point is the effect of manganese on *Aspergillus parasiticus*,[47] and similar effects with yeasts are likely to be found. Irregularities in wall thickness are occasionally noted (Figure 6A) but walls of fairly regular thickness are the norm.

A few species of yeast exhibit hyphae or pseudohyphae (see Volume I, Chapter 2). Certain species of *Endomycopsis, Pichia*, and *Saccharomycopsis* have anastomoses between adjacent hyphae which are apparently induced by minute stalks (denticles).[48] Denticles have been described for only a few species but further studies on their origin, function, and distribution would be of interest.

Bud and birth scars result from cell division. Ultrastructural aspects are discussed in Volume I, Chapter 2 and Volume II, Chapters 3 and 4, while molecular biology of budding and fission are covered in Volume II, Chapters 3 and 4. As discussed in Volume I, Chapter 3, the mechanical and biophysical properties of bud and birth scar regions may be quite different from the remainder of the cell wall.

D. The Periplasmic Space

The periplasmic space is bounded by the plasma membrane and the inner aspect of the cell wall. As an approximation it can be regarded as a coaxial shell whose volume is relatively insensitive to the overall shape of the cell, other things being held constant.[49] However, delineation of the periplasmic space in yeast cross sections is not always obvious. In the cross section depicted in Figure 5A the stained region of the envelope between the cell wall and the plasma membrane, and including invaginations in the

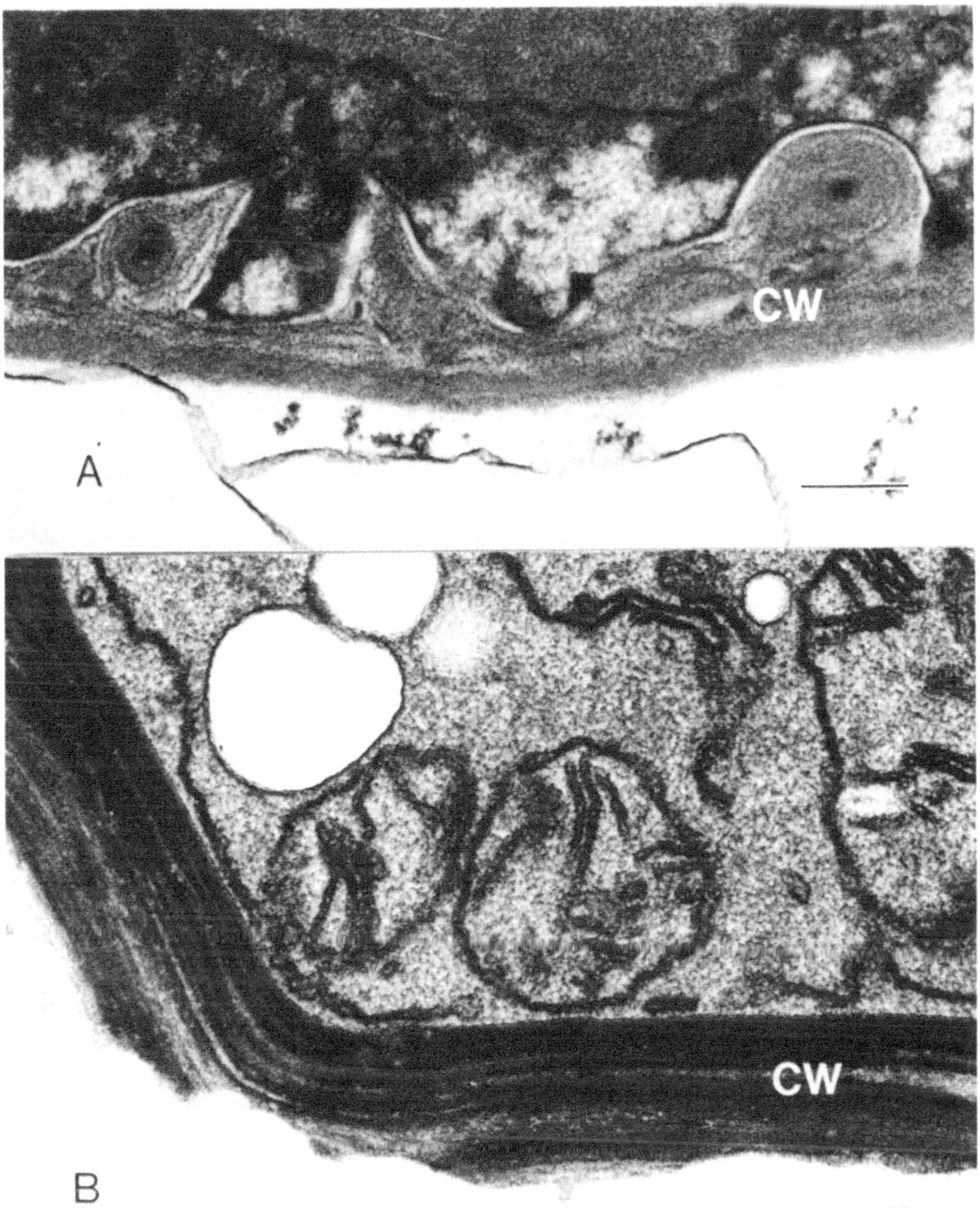

FIGURE 6. Irregular cell walls. (A) *Sporobolomyces roseus*. Note irregular thickening of cell wall (CW). (Glutaraldehyde-osmium tetroxide.) (B) *Sterigmatomyces halophilus*. Note alternate regions of light and dark stained cell wall (CW). (Permanganate) Bars equal 0.25 μm.

latter, can be equated to the periplasmic space. The size of the periplasmic space is greatly increased when cells are placed in hypertonic media (see Volume I, Chapter 3). All of the β-fructofuranosidase and the bulk of the acid phosphatase of *Sacch. cerevisiae* are located in this space (Volume II, Chapter 1). For those cases in which the periplasmic space is stained it is supposedly the glycoprotein constituents that have reacted (Figures 3A and 5B).

Electron micrographs of yeasts occasionally display stained, membrane-circumscribed structures between the plasma membrane and the cell wall. These structures have been called periplasmic bodies. They are especially evident in thin sections of *Sacch. rouxii*,[50] where they are sometimes more prevalent in the bud scar regions of the envelope (Figure 7A). With thin sections of whole cells the relationships among periplasmic bodies, cytoplasm, and the cell wall are difficult to interpret since the plasma membrane is appressed against the inner aspect of the cell wall. Accordingly, a study on periplasmic bodies during protoplast formation from *Sacch. rouxii* cells has been described.[51] The disposition of periplasmic bodies after partial or complete cell wall removal suggests that some may be anatomically free of the protoplasm (Figure 7B). Other periplasmic bodies have a fine pedicel of which the limiting membrane is believed to be continuous with the plasma membrane (Figure 7C). Structures that have been equated to periplasmic bodies are also seen in scanning electron micrographs[51a] of protoplasts of *Sacch. rouxii* (Figure 7D).

The function(s) of periplasmic bodies is unknown. It is postulated that they may be involved in translocation of enzymes destined for the periplasmic space (see Volume II, Chapter 1). Studies with *Sacch. cerevisiae* have shown that periplasmic bodies are rare (or short-lived) in this species under normal physiological conditions, and with identical fixation and processing to that used for *Sacch. rouxii*. On the other hand, similar vesicular structures have been noted on electron images of certain pathogenic fungi which have been challenged with antimycotic drugs.[14] Thus it is possible that periplasmic bodies may also function as a cellular device whereby the fungus can rid itself of denatured plasma membrane. A more precise understanding of periplasmic bodies must await additional studies following isolation and purification.

E. Plasma Membrane

The plasma membrane constitutes the innermost substructural component of the yeast cell envelope. Its preservation in whole yeast cells fixed with glutaraldehyde—osmium tetroxide is often poor. In many yeast species it is more clearly discerned after fixation in permanganate.

The plasma membrane is a tripartite unit membrane about 7.5 nm in thickness. The yeast plasma membrane is characteristically sculptured by occasional invaginations which extend towards the cytoplasm (e.g., Figure 5A). These invaginations are of variable length depending upon the species, age, and growth phase of the cell. Invaginations are also observed in protoplasts (see Figures 7B and 7D). Invaginations are usually not seen in the plasma membrane of young buds. The plasma membrane is strongly stained by the periodate method of Thiéry[39] by dint of its glycolipid or glycoprotein constituents (Figure 5B).

In many yeastlike and filamentous fungi, complex tubular or vesicular structures often occur in the vicinity of the plasma membrane. These structures are variously referred to as lomasomes, plasmalemmasomes, or paramural bodies.[52,53] *Cr. neoformans* seems to display lomasomes by freeze-fracture[37] or thin sections.[38] It has been suggested that lomasomes may perform secretory functions or be involved in formation of cell walls.[54,55]

Electron microscopy of replicas of freeze-fracture surfaces has greatly expanded our knowledge of the yeast cell envelope and of the plasma membrane in particular. These techniques can yield unique information. Figures 8A and 8B illustrate examples of *Sacch. ludwigii* and *Schizosaccharomyces pombe* cell envelopes. The furrowed appearance of the plasma membrane is evident in Figure 8B.

It is now known that frozen plasma membranes fracture along an internal hydrophobic plane to reveal two internal fracture faces.[56] The cytoplasmic half from the

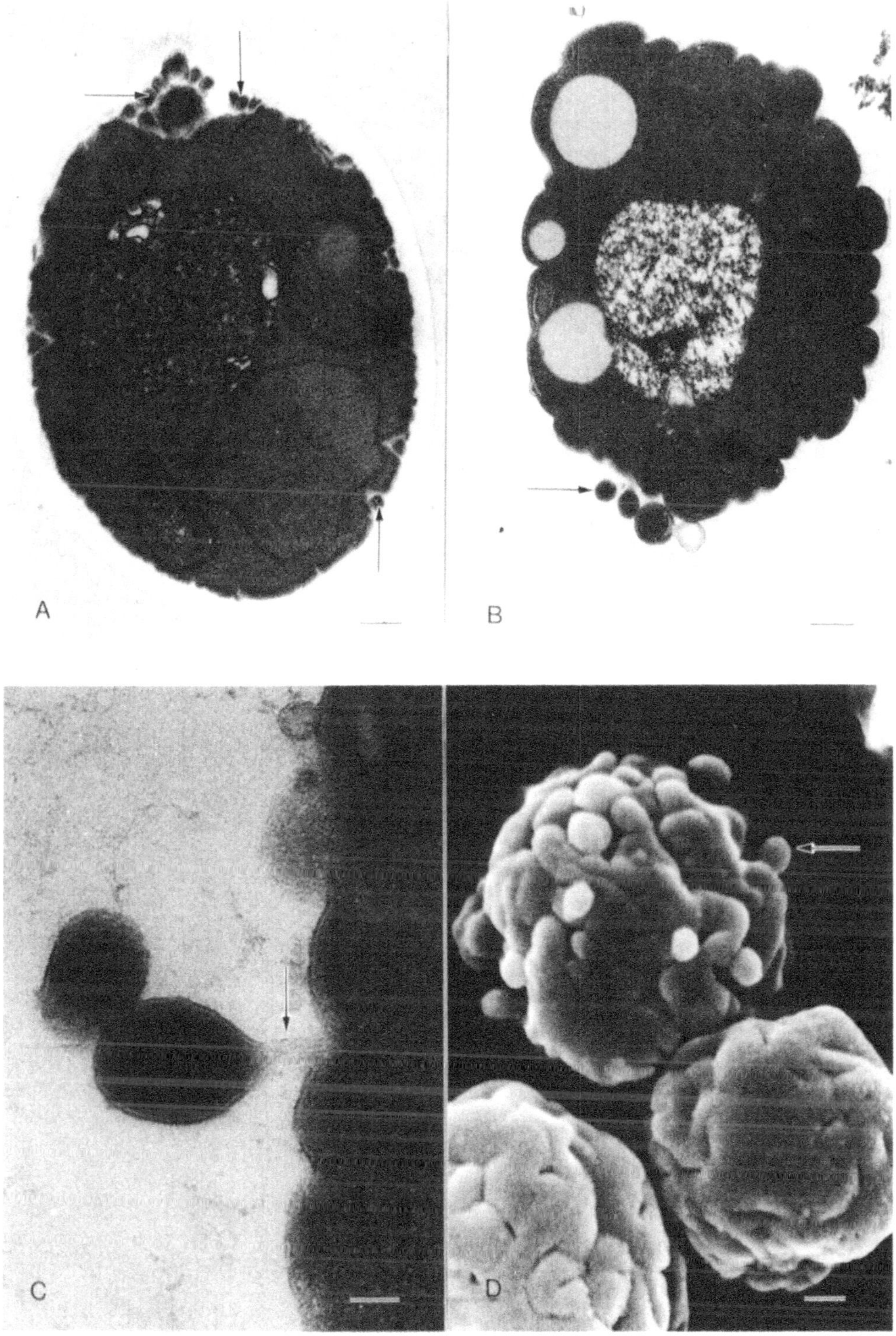

FIGURE 7. Periplasmic bodies in *Saccharomyces rouxii*. All specimens fixed with glutaradehyde—osmium tetroxide. (A) Thin section of young cell. Note periplasmic bodies, some of which are marked by arrows. Bar equals 0.25 μm. (Adapted from Arnold, W. N., Garrison, R. G., and Boyd, K. S., *Appl. Microbiol.*, 28, 1047, 1974). (B) Spheroplast. Bar equals 0.25 μm. (Adapted from Arnold, W. N. and Garrison, R. G., *J. Bacteriol.*, 137, 1386, 1979.) (C) Portion of a protoplast. Note pedicel (arrow) joining periplasmic body to plasma membrane. Bar equals 0.05 μm. (D) Protoplasts by scanning electron microscopy. Bar equals 0.25 μm. (Adapted from Arnold, W. N. and Pringle, A. T., *Curr. Microbiol.*, 3, 283, 1980.)

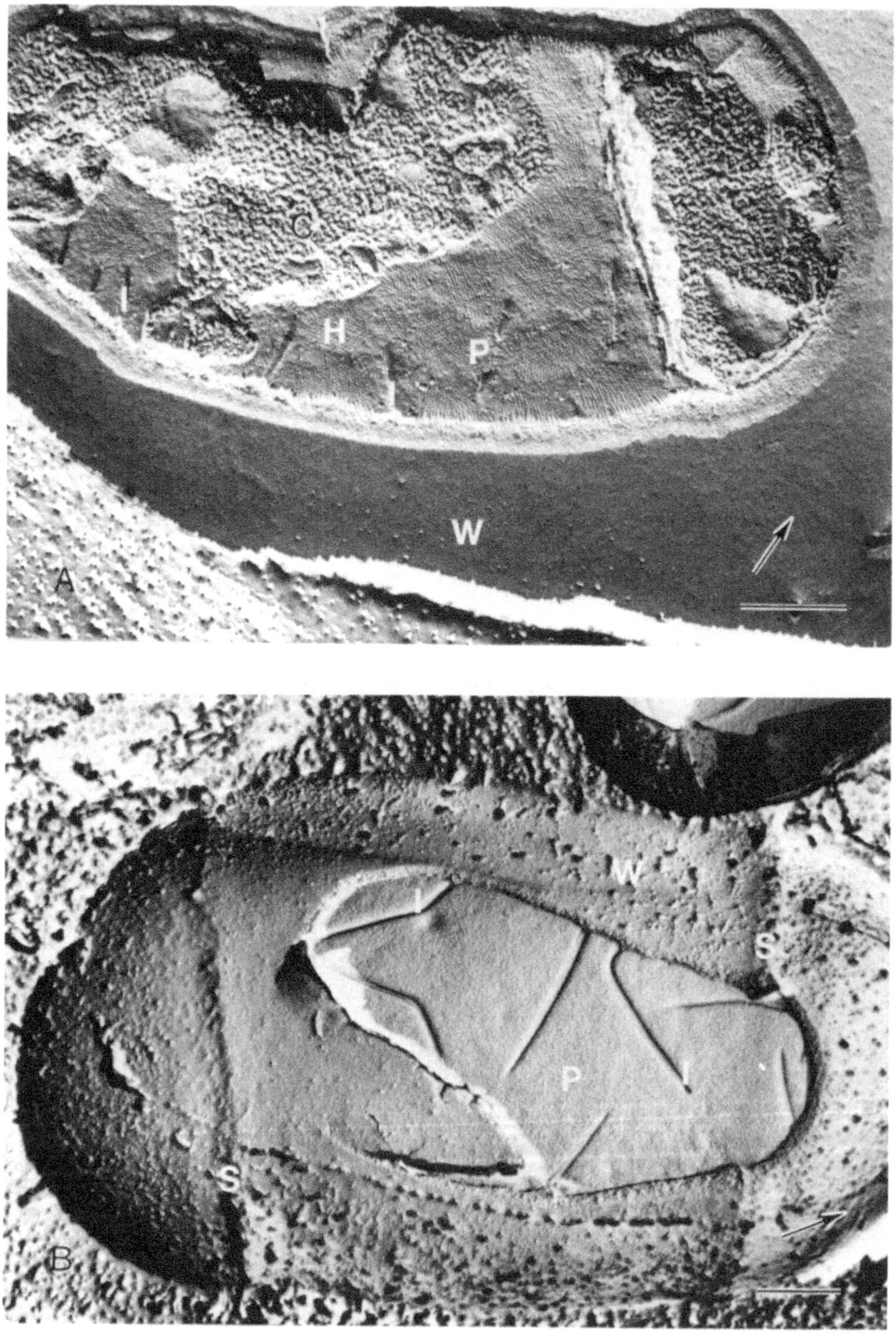

FIGURE 8. Cell envelopes by freeze-fracture. A) Frozen cells of *Saccharomyces cerevisiae* were fractured and then the exposed surface was deeply etched. A carbon replica was prepared and shadowed with metal in the direction indicated by the arrow. The micrograph was obtained by transmission electron microscopy of the replica. Note the cell wall surface (W); P-face of the plasma membrane (P) with some invaginations (I) and hexagonally arrayed particles (H); and the exposed cytoplasm (C). Bar equals 0.5 μm. (Micrograph courtesy of E. Streiblová.) (B) Similarly treated cell of *Schizosaccharomyces pombe:* Note division scars (S); wall surface (W) and P-face of plasma membrane (P) which is furrowed with invaginations (I). Bar equals 0.5 μm. (Micrograph courtesy of E. Streiblová.)

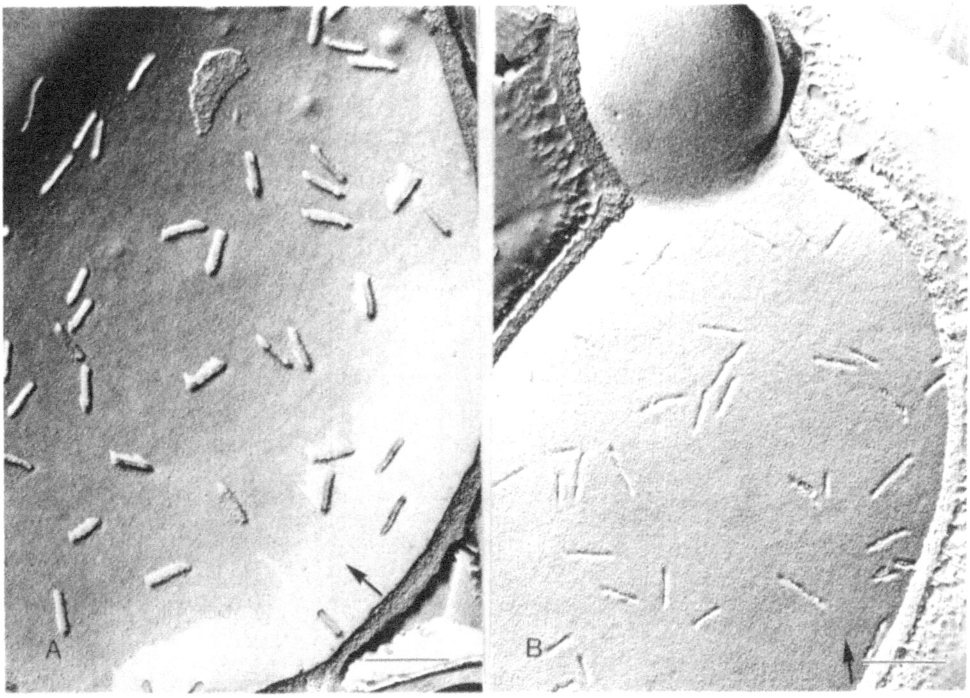

FIGURE 9. Complementary plasma membrane faces of *Candida albicans*. (A) E-face showing numerous protrusion; (B) P-face showing numerous invaginations except in the region corresponding to a young bud. Note that particles are more prevalent on P-face. Bars equal 0.5 μm. (Adapted from Preusser, H. J. and Rostek, H., *Sabouraudia*, 17, 389, 1979.)

original membrane presents the so-called P-face. The outer half of the original membrane presents the so-called E-face. E- and P-faces of a plasma membrane from *C. albicans* are shown in Figure 9. The E-face shows protrusions while the P face shows invaginations. In the intact membrane protrusions and invaginations are of course complementary and constitute the inwardly directed invaginations seen in cross sections and on the surface of protoplasts. Freeze-fractured plasma membranes of *Candida*, *Saccharomyces*, and *Schizosaccharomyces* reveal globular particles which are apparently sandwiched between the two phospholipid layers. These particles tend to be more prevalent on the P-face. Sometimes they occur in hexagonal arrays (Figure 8A). Some preliminary work[57] has appeared on partially purified plasma membranes of *Sacch. cerevisiae*, but unfortunately the material is mostly reformed vesicles which may have suffered some changes in structure.

Yeasts of the *Pityrosporum-Malassezia* group have an unusual cell envelope.[58,59] The inner aspect of the cell wall is sculptured with corrugations (Figure 10A) that are more or less pronounced depending upon cell age and plane of section. The plasma membrane shows corresponding invaginations (Figures 10B and 10C). Of particular interest is the overall spiral arrangement of plasma membrane indentations,[60,61] as shown in Figure 11. This arrangement of grooves is evident even in a young bud (Figure 11A). It has been suggested that the monopolar mode of growth exhibited by *Pityrosporum* species might be related to the unique sculpturing of the plasma membrane.[61]

The most important chemotherapeutic agents used for clinical management of mycotic diseases in man apparently exert their primary effect on the plasma membrane. The polyene macrolides amphotericin B and nystatin, and the imidazole derivatives

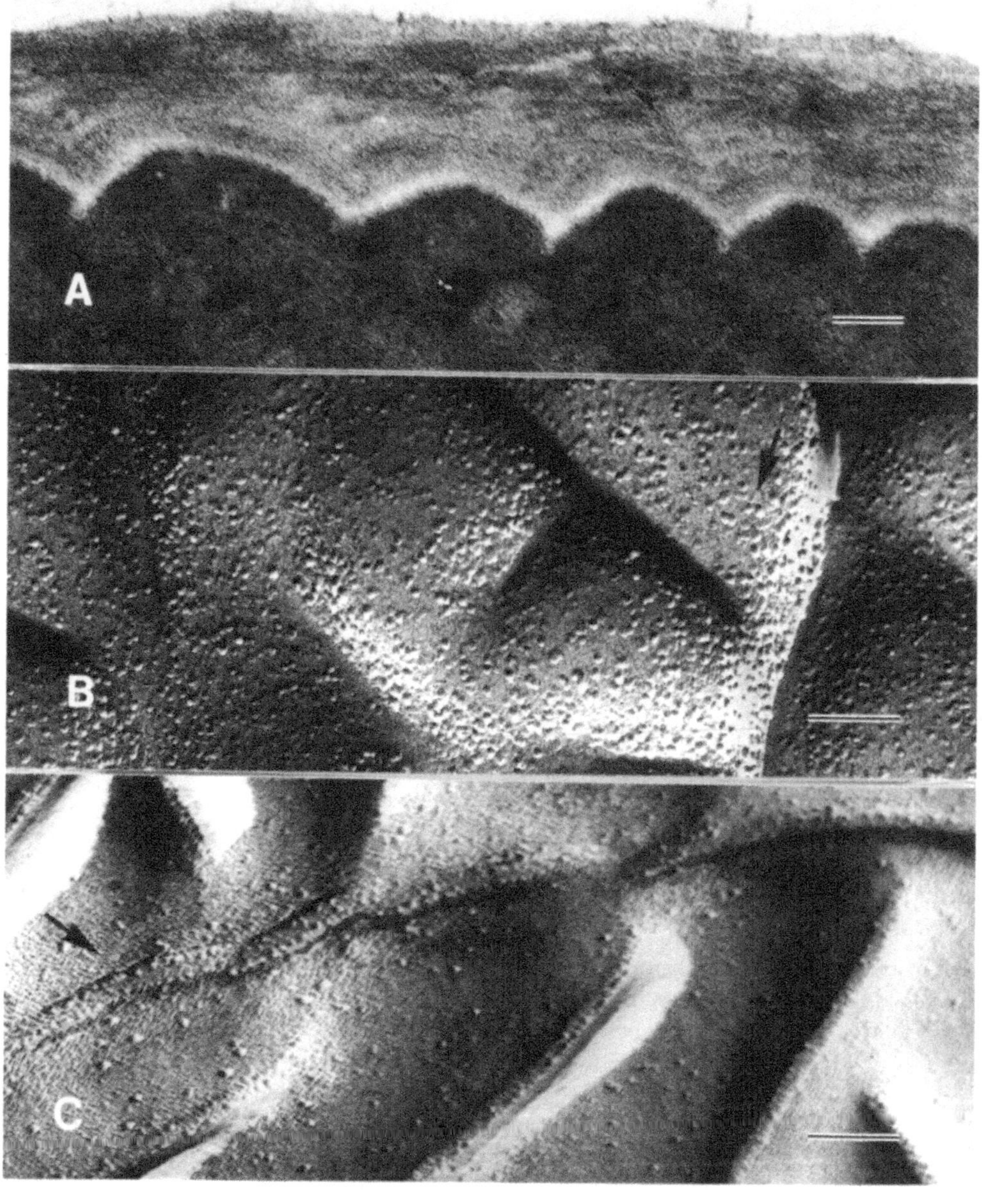

FIGURE 10. Cell envelope of *Pityrosporum orbiculare*. (A) Ultrathin section showing corrugations along inner aspect of cell wall. (Glutaraldehyde—osmium tetroxide.); (B) and (C) freeze-fractured and etched plasma membranes showing (B) invaginations in P-face and (C) protrusions on E-face. Bars equal 0.1 μm. (Micrographs courtesy of A. Breathnach.)

clotrimazole, miconazole, and econazole are the two groups of antimycotic agents in this class.[62] These drugs associate with endogenous sterols and induce membrane denaturation which is revealed by leakage of cytoplasmic constituents (see, for example, Reference 63 and citations therein).

Freeze-fracture studies of the plasma membrane of *C. albicans* after treatment of cells with 22 μ*M* econazole nitrate indicate that morphological alterations occur almost

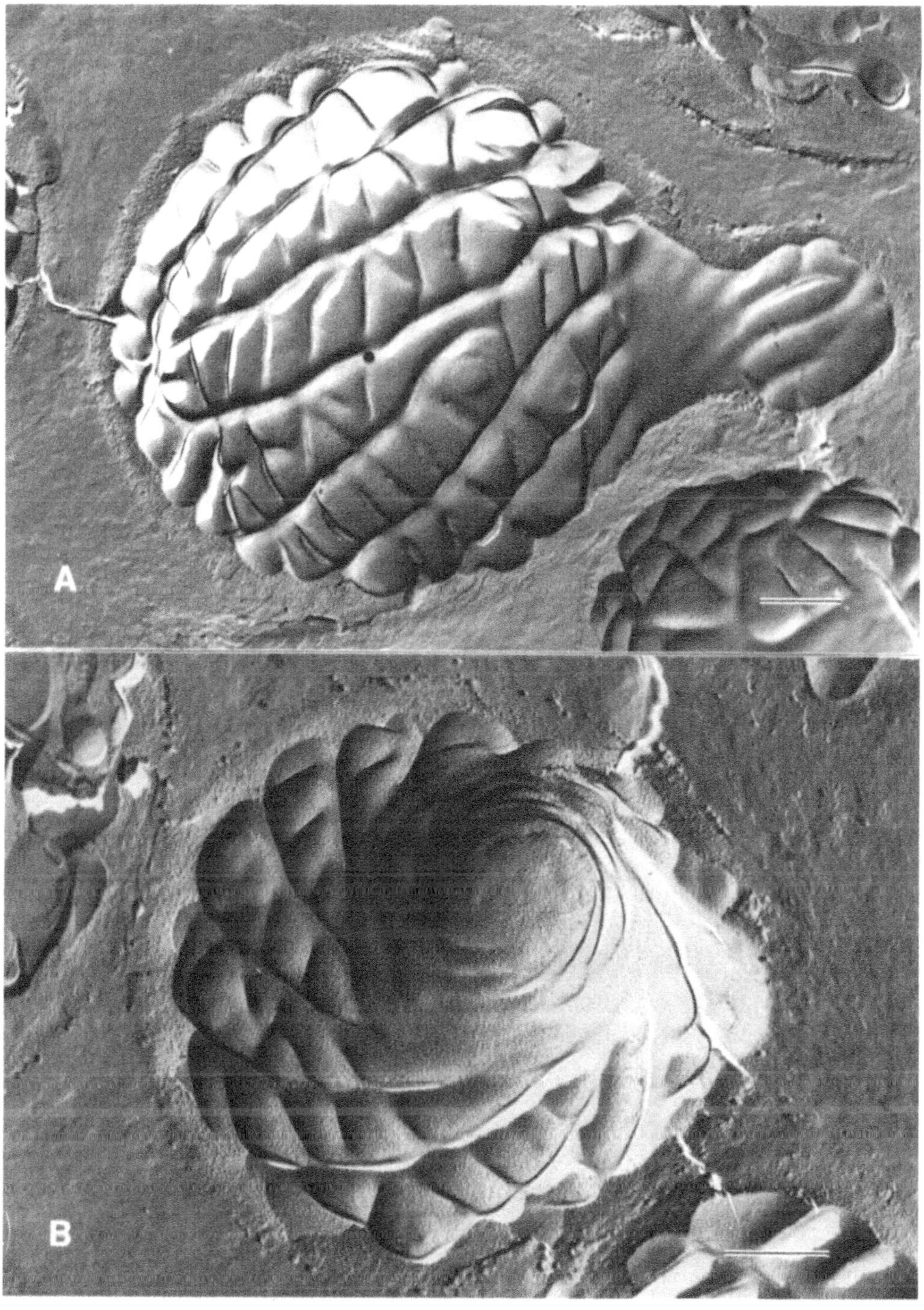

FIGURE 11. Plasma membrane of *Pityrosporum orbiculare*. Freeze-fractured and etched specimens. (A) P-face showing spiral arrangement of invaginations. (Courtesy of A. Breathnach.); (B) E-face showing spiral arrangement of major lines of plasma membrane protrusions. Bars equal 0.5 μm. (Adapted from Breathnach, A.S., Gross, M., and Martin, B., *Sabouraudia*, 14, 105, 1976.)

exclusively on the P-face.[64] The randomly distributed globular particles in the plasma membrane diminish in number per unit area with exposure time (Figure 12). These changes of membrane structure of *C. albicans* were evident on freeze-fracture even when ultrathin sections failed to reveal any membrane alterations.

IV. ELECTRON CYTOCHEMISTRY

Electron cytochemistry includes enzyme cytochemistry, microautoradiography, and electron probe microanalysis. The principles, applications, and limitations of electron cytochemistry have been reviewed by Lewis,[65] and Snitka and Seligman.[66]

Methods developed for mammalian enzyme systems can be adapted to similar investigations of yeasts. Reliable staining methods are currently available for the localization of phosphatases, sulfatases, and haemoproteins but applications in yeast cells may be further complicated by problems of permeation of reagents. One approach to this problem is the cryosectioning of cells into glutaraldehyde.[67,68] However this is of no value for cell envelope enzymes (see Volume II, Chapter 1).

Biochemical and biophysical evidence locates acid phosphatase and β-fructofuranosidase in the periplasmic space of *Saccharomyces* species. At the present time there is no practical method for the cytochemical localization of the latter enzyme. However, methods for locating acid phosphatase activity are available[69] and have been applied to yeasts. The most studied species have been *Sacch. cerevisiae* and *Sacch. rouxii*. Attempts at cytological localization in *Sacch. cerevisiae* by lead capture methods at first yielded conflicting results by placing more evidence on the outer surface of the cell wall,[70-72] but this has been shown to be artifactual (see Volume II, Chapter 1). The true location is the periplasmic space which was previously determined for *Sacch. rouxii*.[72A]

Isotopically labeled substances can be located by mounting the specimen on a suitable support, coating it with an appropriate photographic emulsion, storing it for exposure, developing it, and analyzing the resultant autoradiogram. The studies of Durán et al.[73] on the distribution of chitin synthetase on the yeast plasma membrane exemplify the utility of electron microscopic autoradiography as an indirect means of enzyme localization.

Theoretically, enzyme immunochemistry can be used to specifically detect any cell component that bears antigenic determinants (Volume I, Chapter 6). This approach is restricted to those enzymes which have been purified to homogeneity. Another limitation is that it can only be applied to thin sections since the antibodies will not penetrate the cell wall of intact yeasts.

Electron probe microanalysis is still in the pioneer stage of development. Potentially, it can provide information linking morphology and chemical composition on a subcellular scale. Thin sections of cells can be examined and elements with atomic number of 11 or greater can be detected.

V. CONCLUDING REMARKS

Examinations of thin sections of chemically fixed whole cells and of replicas of freeze-fractured and freeze-etched preparations have yielded a significant body of information on yeast cell envelopes. Yet it is obvious that present techniques of electron microscopy have their limitations. It is disappointing that a species as important as *Sacch. cerevisiae* is refractory to fixation and certainly it cannot be handled in a routine fashion. A number of variables involved in fixation need to be systematically explored. Perhaps a union of yeast microscopists might undertake to share in efforts towards developing improved methodology. This would certainly be worthwhile, because there

FIGURE 12. Effect of econazole nitrate on the plasma membrane of *Candida albicans*. The P-face is shown after (A) 3 hr and (B) 24 hr of treatment. Note the reduction in concentration of particles in (B). Bars equal 0.1 μm. (Adapted from Preusser, H. J., and Rostek, H., *Sabouraudia*, 17, 389, 1979.)

is hardly any aspect of yeast cell biology that could not benefit from companion ultrastructural description.

Recently there have been important technical developments in the scanning transmission electron microscope and the high voltage electron microscope. Both instruments possess unique capabilities but availability is limited at present. Only isolated studies have been reported on aspects of fungal cytology as observed in electron microscopes operating within the megavolt range.[74-77] Ultimately it may be possible to study the living yeast cell at high magnification with this class of electron microscope.[75] The implications are exciting.

ACKNOWLEDGMENTS

I am very grateful to investigators who contributed electron micrographs. Karen Boyd Grantham provided excellent technical assistance on micrographs taken at the Veterans Administration Hospital.

REFERENCES

1. **Matile, P., Moor, H., and Robinow, C. F.,** Yeast cytology, in *The Yeasts*, Vol. 3, Rose, A. H. and Harrison, J. S., Eds., Academic Press, New York, 1969, 220.
2. **Hayat, M. A.,** *Principles and Techniques of Electron Microscopy: Biological Applications*, Vol. 1, Van Nostrand Reinhold, New York, 1970.
3. **Hayat, M. A.,** *Basic Electron Microscopic Techniques*, Van Nostrand Reinhold, New York, 1972.
4. **Pease, D. C.,** *Histological Techniques for Electron Microscopy*, 2nd ed., Academic Press, New York, 1964.
5. **Sjöstrand, F. S.,** *Electron Microscopy of Cells and Tissues*, Vol. 1, Academic Press, New York, 1967.
6. **Greenhalgh, G. N. and Evans, L. V.,** Electron microscopy, in *Methods in Microbiology*, Vol. 4, Booth, C., Ed., Academic Press, New York, 1971, 517.
7. **Kopp, F.,** Electron microscopy of yeasts, *Methods Cell Biol.*, 12, 23, 1975.
8. **Moor, H. and Mühlethaler, K.,** Fine structure in frozen-etched yeast cells, *J. Cell Biol.*, 17, 609, 1963.
9. **Bullivant, S.,** Freeze-etching and freeze-fracture, in *Advanced Techniques in Biological Electron Microscopy*, Koehler, J. K., Ed., Springer-Verlag, New York, 1973, 67.
10. **Koehler, J. K.,** The freeze-fracture technique, in *Principles and Techniques of Electron Microscopy*, Vol. 2, Hayat, M. A., Ed., Van Nostrand Reinhold, New York, 1962, 53.
11. **Rash, J. E. and Hudson, C. S.,** *Freeze-fracture, Methods, Artifacts, and Interpretations*, Raven Press, New York, 1979.
12. **Stolinski, C. and Breathnach, A. S.,** *Freeze-Fracture Replication of Biological Tissues. Techniques, Interpretation and Application*, Academic Press, New York, 1975.
13. **Hoch, H. C.,** Use of permanganate to increase electron opacity of fungal walls, *Mycologia*, 69, 1209, 1977.
14. **Garrison, R. G.,** unpublished observations.
15. **Mollenhauer, H. H.,** Permanganate fixation of plant cells, *Biophys. Biochem. Cytol.* 6, 431, 1959.
16. **Persi, M. A. and Burnham, J. C.,** Use of tannic acid as a fixative-mordant to improve the ultrastructural appearance of *Candida albicans* blastospores, *Sabouraudia*, 19, 1, 1981.
17. **Djaczenko, W. and Cassone, A.,** Visualization of new ultrastructural components in the cell wall of *Candida albicans* with fixatives containing TAPO, *J. Cell Biol.*, 52, 186, 1972.
18. **Cassone, A., Simonetti, N., and Strippoli, V.,** Wall structure and bud formation in *Cryptococcus neoformans, Arch. Microbiol.*, 95, 205, 1974.
19. **Cassone, A.,** Improved visualization of wall ultrastructure in *Saccharomyces cerevisiae, Experientia*, 29, 1303, 1973.
20. **May, R.,** Licht- und Electronenoptische Untersuchungen an Protein-Kristallen und Mikrotubuli an Hefeprotoplasten, *Z. Allg. Mikrobiol.*, 14, 409, 1974.

21. **Glauert, A. M.**, The fixation and embedding of biological specimens, in *Techniques for Electron Microscopy*, F. A. Davis, Philadelphia, 1967.
22. **Furtado, J. S.**, The fibrin clot; a medium for supporting loose cells and delicate structures during processing for microscopy, *Stain Technol.*, 45, 19, 1970.
23. **Hubert, M. T., Carasso, N., and Favard, P.**, Methods de rassemblement de petits organismes en suspension pour l'inclusion dans les milieux visquex, *J. Microscopie*, 1, 163, 1962.
24. **Sawicki, W. and Lipetz, J.**, Albumin embedding and individual mounting of one or many mammalian ova on slides for fluid processing, *Stain Technol.*, 46, 261, 1971.
25. **Shands, J. W.**, Embedding free-floating cells and microscopic particles; serum albumin coagulum-epoxy resin, *Stain Technol.*, 43, 15, 1968.
26. **Reedy, M.**, Section staining for electron microscopy, *J. Cell Biol.*, 26, 309, 1965.
27. **Soloff, B.**, Buffered potassium permanganate-uranyl acetate-lead citrate staining sequence for ultrathin sections, *Stain Technol.*, 48, 159, 1973.
28. **Spurr, A. R.**, A low viscosity epoxy resin embedding medium for electron microscopy, *J. Ultrastruct. Res.*, 26, 31, 1960.
29. **Poon, N. H. and Day, A. W.**, Fungal fimbriae. I. Structure, origin, and synthesis, *Can. J. Microbiol.*, 21, 537, 1975.
30. **Day, A. W. and Poon, N. H.**, Fungal fimbriae. II. Their role in conjugation in *Ustilago violacea, Can. J. Microbiol.*, 21, 547, 1975.
31. **Day, A. W., Poon, N. H., and Stewart, G. C.**, Fungal fimbriae. III. The effect on flocculation in *Saccharomyces, Can. J. Microbiol.*, 21, 558, 1975.
32. **Phaff, H. J.**, Structure and biosynthesis of the yeast cell envelope, in *The Yeasts,* Vol. 2, Rose, A. H. and Harrison, J. S., Eds., Academic Press, New York, 1971, 135.
33. **Luft, J. H.**, Electron microscopy of cell extraneous coats as revealed by ruthenium red staining, *J. Cell Biol.*, 23, 54A, 1964.
34. **Luft, J. H.**, Ruthenium red and violet. I. Chemistry, purification, methods of use for electron microscopy and mechanisms of action, *Anat. Rec.*, 171, 347, 1971.
35. **Martin, B. J. and Spicer, S. S.**, Concanavalin A-iron dextran technique for staining cell surface mucosubstances, *J. Histochem. Cytochem.*, 22, 206, 1974.
36. **Mahvi, T. A., Spicer, S. S., and Wright, N. J.**, Cytochemistry of acid mucosubstances and acid phosphatase in *Cryptococcus neoformans, Can. J. Microbiol.*, 20, 833, 1974.
37. **Takeo, K., Uesaka, I., Uehira, K., and Nishiura, M.**, Fine structure of *Cryptococcus neoformans* grown in vitro as observed by freeze-etching, *J. Bacteriol.*, 113, 1442, 1973.
38. **Edwards, M. R., Gordon, M. A., Lapa, E. W., and Ghiorse, W. C.**, Micromorphology of *Cryptococcus neoformans, J. Bacteriol.*, 94, 496, 1967.
39. **Thiéry, J. P.**, Mise en evidence des polysaccharides sur coupes fines en microscopie electronique, *J. Microscopie*, 6, 987, 1967.
40. **Cassone, A., Mattia, E., and Boldrini, L.**, Agglutination of blastospores of *Candida albicans* by concanavalin A and its relationship with the distribution of mannan polymers and the ultrastructure of the cell wall, *J. Gen. Microbiol.*, 105, 263, 1978.
41. **Kreger-van Rij, N. J. W. and Veenhuis, M.**, A comparative study of the cell wall structure of basidiomycetous and related yeasts, *J. Gen. Microbiol.* 68, 87, 1971.
42. **Ballou, C.**, Structure and biosynthesis of the mannan components of the yeast cell envelope, *Adv. Microb. Physiol.*, 14, 93, 1976.
43. **Kerridge, D., Koh, T. Y., Marriott, M. S., and Gale, E. F.**, The production and properties of protoplasts from the dimorphic yeast *Candida albicans*, in *Microbial and Plant Protoplasts*, Perberdy, J. F., Rose, A. H., Rodgers, H. J., and Cocking, E, C., Eds., Academic Press, New York, 1976, 23.
44. **Chattaway, F. W., Shenolikar, S., O'Reilly, J., and Barlow, A. J. E.**, Changes in the cell surface of the dimorphic forms of *Candida albicans* by treatment with hydrolytic enzymes, *J. Gen. Microbiol.*, 95, 335, 1976.
45. **Hagedorn, H.**, Die Feinstruktur der Hefezellen. I. Zellwand, Sporen, Cytoplasma, Endoplasmatisches Retikulum, *Protoplasma*, 58, 250, 1964.
46. **McMurrough, I. and Rose, A. H.**, Effect of growth rate and substrate limitation on the composition and structure of the cell wall of *Saccharomyces cerevisiae, Biochem. J.*, 105, 189, 1967.
47. **Garrison, R.G., and Boyd, K. S.**, Ultrastructural studies of induced morphogenesis by *Aspergillus parasitious, Sabouraudia*, 12, 179, 1971.
48. **Kreger-van Rij N. J. W., and Veenhuis, M.**, Electron microscopy of some special cell contacts in yeasts, *J. Bacteriol.*, 113, 350, 1973.
49. **Arnold, W. N.**, Volume and enzyme content of the periplasmic space in yeast, *Physiol. Chem. Physics*, 5, 117, 1973.
50. **Arnold, W. N., Garrison, R. G., and Boyd, K. S.**, Periplasmic structure in *Saccharomyces rouxii* (Boutroux), an osmophile, *Appl. Microbiol.*, 28, 1047, 1974.

51. **Arnold, W. N. and Garrison, R. G.,** Isolation and characterization of protoplasts from *Saccharomyces rouxii, J. Bacteriol.*, 137, 1386, 1979.
51a. **Arnold, W. N. and Pringle, A. T.,** Scanning electron microscopy of cells and protoplasts of *Saccharomyces rouxii, Curr. Microbiol.*, 3, 283, 1980.
52. **Moore, R. T. and McAlear, J. H.,** Fine structure of mycota. V. Lomasomes-previously uncharacterized hyphal structures, *Mycologia*, 53, 194, 1968.
53. **Marchant, R. and Moore, R. T.,** Lomasomes and plasmalemmasomes in fungi, *Protoplasma*, 76, 235, 1973.
54. **Kozar, F. and Weijer, J.,** Electron-dense structures in *Neurospora crassa*. II. Lomasome-like structures, *Can. J. Genet. Cytol.*, 11, 617, 1969.
55. **Wilsenach, R. and Kessel, M.,** The role of lomasomes in wall formation in *Penicillium vermiculatum, J. Gen. Microbiol.*, 40, 401, 1965.
56. **Branton, D.,** Freeze-etching studies of membrane structure, *Philos. Trans. R. Soc.* London Ser B, 261, 133, 1971.
57. **Matile, P.,** Properties of the purified cytoplasmic membrane of yeasts, *FEBS Symp.*, 20, 39, 1970.
58. **Keddie, F. M. and Barajas, L.,** Three-dimensional reconstruction of *Pityrosporum* yeast cells based on serial section electron microscopy, *J. Ultrastruct. Res.*, 29, 260, 1969.
59. **Kreger-van Rij, N. J. W. and Veenhuis, M.,** An electron microscopic study of the yeast *Pityrosporum ovale, Arch. Mikrobiol.*, 71, 123, 1970.
60. **Breathnach, A. S., Gross, M., and Martin, B.,** Freeze-fracture replication of cultural *Pityrosporum orbiculare, Sabouraudia*, 14, 105, 1976.
61. **Takeo, K. and Nakai, E.,** The plasma membrane of *Pityrosporum* has natural markers to show how the cell grows, *9th Int. Cong. Elect. Micros.*, 2, 404, 1978.
62. **Kobayashi, G. S. and Medoff, G.,** Antifungal agents: Recent developments, *Annu. Rev. Microbiol.*, 31, 291, 1977.
63. **Arnold, W. N., Pringle, A., and Garrison, R. G.,** Amphotericin B-induced changes in K^+ content, viability, and ultrastructure of yeast-phase *Histoplasma capsulatum, J. Bacteriol.*, 141, 350, 1980.
64. **Preusser, H. J. and Rostek, H.,** Freeze-fracture studies of the plasmalemma of *Candida albicans* after treatment with econazole nitrate, *Sabouraudia*, 17, 389, 1979.
65. **Lewis, P. R.,** Electron microscopical localization of enzymes, in *Advances in Optical and Electron Microscopy,* Vol. 6, Barer, R. and Casslett, E. B., Eds., Academic Press, New York, 1975, 171.
66. **Snitka, T. K. and Seligman. A. M.,** Ultrastructural localization of enzymes, *Annu. Rev. Biochem.*, 40, 375, 1971.
67. **Borgers, M. and De Nollin, S.,** The preservation of subcellular organelles of *Candida albicans* using conventional fixatives, *J. Cell Biol.*, 62, 574, 1974.
68. **De Nollin, S., Throné, F., and Borgers, M.,** Electron cytochemistry of *Candida albicans, J. Histochem. Cytochem.*, 23, 758, 1975.
69. **Essner, E.,** Phosphatases, in *Electron Microscopy of Enzymes, Principles and Methods*, Vol. 1, Hayat, M. A., Ed., Van Nostrand Reinhold, New York, 1973, 44.
70. **Bauer, H. and Sigarlarkie, E.,** Cytochemistry on ultrathin frozen sections of yeast cells, *J. Microscopie*, 99, 205, 1973.
71. **Günther, T., Kettner, W., and Merker, H. J.,** Über das Verholten und die Lokalisation der Sauren Phosphatase von Hefezellen bei Repression und Derepression, *Exp. Cell Res.*, 45, 133, 1966.
72. **Linnemans, W. A. M., Boer, P., and Elbers, P. F.,** Localization of acid phosphatase in *Saccharomyces cerevisiae*: A clue to cell wall formation, *J. Bacteriol.*, 131, 638, 1977.
72a. **Arnold, W. N. and Garrison, R. G.,** An Fe^{3+}-activated acid phosphatase in *Saccharomyces rouxii, J. Biol. Chem.*, 254, 4919, 1979.
73. **Duran, A., Cabib, E., and Bowers, B.,** Chitin synthetase distribution on the yeast plasma membrane, *Science*, 203, 363, 1979.
74. **Dupouy, G., Perrier, F., and Durrieu, L.,** L'observation de matière vivante au noyen d'un microscope électronique fonctionment sous haute tension, *C. R. Acad. Sci.*, 251, 2856, 1960.
75. **Dupouy, G.,** Advantages of megavolt electron microscopy in biological research, *Ultramicroscopy*, 2, 199, 1977.
76. **Crang, R. F. and Pechak, D. G.,** Serial section reconstruction of black yeast, *Aureobasidium pullularis*, by means of high voltage electron microscopy, *Protoplasma*, 96, 225, 1978.
77. **Harris, J. L.,** Some three-dimensional aspects of *Ceratocystis ulmi* as observed by high-voltage electron microscopy, *Mycologia*, 67, 332, 1975.

SUBJECT INDEX

A

B

C

D

H

I

L

M

N

O

P

R

S

T

U

V

W

X

Y

Z

TAXONOMIC INDEX*

* Yeast and yeast-like species appear in boldface italics. All other species appear in italics.